T0135876

Matthias Heß

Analysis of the Navier-Stokes Equations for Geophysical Boundary Layers

Bibliografische Information der Deutschen Nationalbibliothek

Die Deutsche Nationalbibliothek verzeichnet diese Publikation in der Deutschen Nationalbibliografie; detaillierte bibliografische Daten sind im Internet über http://dnb.d-nb.de abrufbar.

Dissertation
TU Darmstadt
D 17

ISBN 978-3-8325-2304-6

Logos Verlag Berlin GmbH
Comeniushof, Gubener Str. 47,
10243 Berlin
Tel.: +49 (0)30 42 85 10 90
Fax: +49 (0)30 42 85 10 92
INTERNET: http://www.logos-verlag.de

Contents

Introduction

In this thesis we will consider problems in the framework of the Navier-Stokes equations, which are the fundamental mathematical description of the hydrodynamics of incompressible viscous fluids (also called *Newtonian fluids*). In principal, for $n \geq 2$ they are given by

$$\left\{\begin{array}{rcl} \partial_t u - \nu\Delta u + (u \cdot \nabla)u + \nabla p &=& 0, \quad t \in (0,T),\ x \in \Omega, \\ \operatorname{div} u &=& 0, \quad t \in (0,T),\ x \in \Omega, \\ u(t,x) &=& 0, \quad t \in (0,T),\ x \in \partial\Omega, \\ u(0,x) &=& u_0, \quad x \in \Omega, \end{array}\right. \tag{1}$$

where $(0,T)$ for $0 < T \leq \infty$ is some time interval, $\Omega \subset \mathbb{R}^n$ is a domain, and $\nu > 0$ denotes the viscosity of the fluid. The function $u : [0,T) \times \Omega \to \mathbb{R}^n$ describes the velocity of the fluid and $p : [0,T) \times \Omega \to \mathbb{R}$ is the corresponding pressure. The Navier-Stokes equations serve as the basic model for various disciplines of engineering and natural science concerning fluid mechanics. For example, the flow of oil in a pipeline or the streaming of water in an ocean may be described by system (1). Since air may be assumed to behave as a Newtonian fluid for speed of movements far below the sonic velocity, the Navier-Stokes equations are also deployed for the modeling of atmospheric phenomena.

System (1) has been being object of extensive mathematical investigations for the last decades. Starting from the pathbreaking work of Leray [Ler34], the Navier-Stokes equations were tackled by many authors. For example, in the articles [KL63], [FK64], and [Sol77] the authors developed several approaches to the existence of strong solutions. Kato proved in [Kat84] the existence of unique strong local-in-time solutions of (1) for $u_0 \subset L^n_\sigma(\Omega)$. In the case of small initial data u_0 he even obtained global solutions. Nevertheless, there still remain a lot of open questions, in particular the uniqueness of global weak solutions or the existence of global strong solution for general initial data and dimension $n \geq 3$. For a comprehensive overview of the mathematical theory of the Navier-Stokes equations we refer to the monographs [Tem77], [Gal94a] and [Gal94b], or [Soh01].

A frequently used strategy in the context of the nonlinear Navier-Stokes equations is to consider the linearized system, the so-called Stokes equations, which are obtained from (1) by dropping the convective term $(u \cdot \nabla)u$. A deep knowledge of the linearized equations has proved to be helpful for the treatment of the nonlinear system. In particular, let Ω be a domain, such that the Helmholtz projection P exists. Applying P to (1), we obtain the evolution equation

$$\begin{cases} u'(t) + \nu A u(t) &= -P(u(t) \cdot \nabla)u(t), \quad t > 0, \\ u(0) &= u_0, \end{cases} \tag{2}$$

in $L^p_\sigma(\Omega)$, where $A = -P\Delta$ denotes the Stokes operator in $L^p_\sigma(\Omega)$.

The representation of the Navier-Stokes equations in the form of an evolution equation in the Banach space $L^p_\sigma(\Omega)$ provides the starting point for employing operator theory for further investigations. In this context, detailed information about the spectral properties of the corresponding linear operator is of particular interest. In this thesis we present a proof that for bounded or exterior domains Ω with sufficiently smooth boundary the Stokes operator is $\mathcal{R}$-sectorial with angle 0. In particular, this implies that the Stokes equations have maximal regularity. The latter fact has already been proved by Solonnikov using potential theoretic arguments (see [Sol77]). Our proof mainly relies on operator theory. Moreover, by an application of the Bogovskiĭ operator, we show that the usual localization procedure for elliptic problems can be transferred to Stokes equation.

As mentioned before, the Navier-Stokes equations are of fundamental importance for the mathematical description of phenomena concerning fluid dynamics. The pure Navier-Stokes equations model the inertia and the frictional forces occuring in a moving fluid. In applications, several other forces may be incorporated. For example, in the modeling of geophysical flows one must take into account the earth's rotation. Considering a moving mass in a rotating framework, there are several forces acting. In the mathematical modeling, the most of these forces may be absorbed in the pressure term of the Navier-Stokes equations. This does not work for the so-called Coriolis force. If we consider the motion of a fluid in a three-dimensional domain, whose velocity in a point x is given by $u(x)$, then the Coriolis force acting in x is described by the term $\mathbf{e} \times u$, where $\mathbf{e}$ is the axis of rotation. Hence, adding this term to the left-hand side of system (1), we obtain a model, which is applicable for the description of geophysical flows, e.g. of water in an ocean or of atmospheric streamings.

For Ω being the three-dimensional halfspace $\mathbb{R}^3_+$, the Navier-Stokes equations with Coriolis force term have a famous stationary (i.e. time-independent) solution, which is explicitly given, the so-called Ekman spiral (see [Ekm05]). In natural science, this solution serves as a well accepted model for geostrophic flows in oceans or in the atmosphere. Hence, its mathematical understanding is of significant importance for applications. In particular, the stability of the time-independent solution under perturbation is of interest. Here, we talk of stability of dynamical systems in the sense of Lyapunov. To be more precise, let u_E denote the Ekman spiral and consider a time-dependent solution $u + u_E$ of the Navier-Stokes equations with Coriolis force term with a perturbation $u : [0, \infty) \to X$ for some Banach space X. Then we say the Ekman spiral is stable, if u remains bounded. Otherwise we say that it is unstable. Any hydrodynamical system is characterized by some non-dimensional characteristic, the so-called Reynolds number Re. This ratio may be considered as a measure to what extent the system tends to develop turbulences. Hence, it seems natural to conjecture that there is a critical value Re_c with the property that the corresponding nonlinear flow is stable, if $\mathrm{Re} < \mathrm{Re}_c$, and that it is unstable for $\mathrm{Re} > \mathrm{Re}_c$. In fact, using numerical simulations, Lilly proved in [Lil66] that the Ekman spiral becomes unstable in case of a Reynolds number Re, which exceeds 55. In [DDG99] it is shown that for small Reynolds numbers a perturbation of the Ekman spiral remains bounded in the L^2-norm. In this thesis we will give a rigorous proof of the fact that the Ekman spiral is asymptotically stable for L^2-perurbations in case of Reynolds numbers, that are sufficiently small. In particular, in this case we even obtain a polynomial decay of the perturbation. Since our result holds for values below $\mathrm{Re} \approx 1$, there remains a gap in the range of Reynolds numbers, where the stability behaviour of the Ekman spiral is not known.

In applications, often the underlying domain is not chosen to be a halfspace. Actually, many natural archetyps like oceans and the troposphere are better met in a modeling by an infinite layer. The Stokes equations on such domains are understood very well. In fact, in the works of Abels (see [Abe05a], [Abe05b]) as well as in the ones of Abe and Shibata ([AS03a], [AS03b]) detailed descriptions of the spectral properties of the Stokes operator in infinite layers are given. In particular, these results yield that the negative Stokes operator in these domains is the generator of an analytic semigroup, which is exponentially stable. By the construction of a stationary solution of the Navier-Stokes equations with Coriolis force term in infinite layers, it is possible to adapt the theory of the Ekman spiral to such domains. Then the same questions concerning stability as in the halfspace case arise. Since the corresponding linear operator is closely related to the Stokes operator, we may transfer the exponential decay of the semigroup and obtain exponential stability of the Ekman spiral in infinite layers.

The structure of the present thesis is as follows.

In Chapter 1 we will give a survey of the basic notations and mathematical concepts used in the course of this thesis.

Chapter 2 collects certain results on the Stokes equations in the halfspace, which will be used in the following chapters. In particular, we will state the property of the Stokes operator to be $\mathcal{R}$-sectorial in $L^p_\sigma(\Omega)$ for $1 < p < \infty$, which follows from a result in [DHP01], and give a corresponding estimate on the $\mathcal{R}$-bound of its resolvent mapping from $L^p_\sigma(\Omega)$ to $W^{s,p}(\Omega)$ for $s \in [0,2]$. Moreover, we prove that a result given in [NS03] concerning the decay of the pressure p in the Stokes equations can be transferred to the context of $\mathcal{R}$-boundedness.

Chapter 3 is devoted to the question, whether the Stokes equations in a bounded or exterior domain Ω have maximal regularity. By the characterization theorem maximal regularity of Weis ([Wei01]), this problem reduces to the question, if the Stokes operator in $L^p_\sigma(\Omega)$ is $\mathcal{R}$-sectorial. Considering a bounded domain Ω with sufficiently smooth boundary, by a localization method this problem may be traced back to the halfspace case. Then the result on the $\mathcal{R}$-sectoriality of the Stokes operator in $\mathbb{R}^n_+$ can be transferred to Ω. We remark, that the usual localization procedure for elliptic problems is not applicable in this case because it does not respect the divergence free condition. We handle this problem using the Bogovskiĭ operator. Roughly speaking, this operator acts as a right inverse of the divergence operator. Its application allows to obtain a solution to the Stokes resolvent equation and $\mathcal{R}$-bounds of the corresponding solution operator. The exterior domain case will then be handled by a combination of the result for bounded domains and the corresponding properties of the Laplacian in $L^p(\mathbb{R}^n)$.

In Chapter 4 we present an existence result for weak L^2-solutions to the Navier-Stokes equations with an additive linear perturbation B in a domain Ω being particularly the halfspace or an infinite layer. Here the mentioned perturbation has to fulfill several conditions. To be more precise, the key for the proof is that the operator $-(A + PB)$ is the generator of a semigroup of contractions in $L^2_\sigma(\Omega)$. We remark, that this condition is satisfied for example for $Bu := \mathbf{e} \times u$ being the Coriolis force term, which appears, if we consider the Navier-Stokes equations in a rotating framework. Herewith, we generalize a similar result of Miyakawa and Sohr (see [MS88]) for the case of the pure Navier-Stokes equations. Actually, in their proof they used the property of the negative Stokes operator to be selfadjoint in $L^2_\sigma(\Omega)$ and to generate a bounded analytic semigroup. We are able to modify these conditions. In particular, even in the case $Bu = \mathbf{e} \times u$, the operator $-(A + PB)$ does not generate a bounded analytic semigroup.

Chapter 5 deals with the stability of the Ekman spiral in $\mathbb{R}^3_+$. Considering a perturbation v of u_E, we find that v has to be a solution to some perturbed Navier-Stokes system. If the Reynolds number corresponding to the considered system is small enough, we may show that the associated linear operator is the generator of a semigroup of contractions. Hence, applying the existence result given in Chapter 4, we are able to construct a global-in-time weak L^2-solution v to this system. In the following we show that $\lim_{t\to\infty} \|v\|_2 = 0$ holds. If the semigroup generated by the operator associated to the perturbed Navier-Stokes equations applied to the initial value v_0 admits a polynomial decay, then the solution also does. The proof that the mentioned semigroup is strongly stable will be the essential step for showing these estimates.

In Chapter 6 the Ekman spiral on infinite layers Ω_b is being investigated. The idea to transfer the Ekman spiral to such domains is inspired by the corresponding results concerning the Stokes operator. In particular, by the results of Abels ([Abe05a]) as well as the ones of Abe and Shibata ([AS03b]) we know that the negative Stokes operator on $L^p_\sigma(\Omega_b)$ generates an exponentially decaying semigroup for $1 < p < \infty$. Similar to the halfspace case, for small Reynolds numbers we may construct global-in-time weak L^2-solutions to the perturbed equation. In fact, we are able to prove that such solutions admit an exponential decay. Furthermore, using perturbation theory we prove that even the negative Stokes-Coriolis-Ekman operator in $L^p_\sigma(\Omega_b)$ is the generator of an exponentially decaying semigroup for $1 < p < \infty$. Adopting Kato's iteration method, we are able to prove the existence of strong L^3-solutions of the perturbed equation, which exist locally in time, in general. For sufficiently small initial data we even show that this strong solution is a global one. In this case the exponential decay of the semigroup transfers to the solution. Hence, we obtain exponential stability of the Ekman spiral.

Acknowledgements

First of all, I would like to thank my advisor Prof. Matthias Hieber for the chance to work on these interesting problems. His guidance and many fruitful suggestions contributed invaluably to the development of this thesis.

I also would like to thank Prof. Reinhard Farwig for acting as co-referee.

I wish to thank my colleagues Matthias Geißert, Robert Haller-Dintelmann, and Horst Heck for many interesting discussions and their persistent support. They were always prepared to answer my questions patiently and helped to improve this thesis a lot. In particular, I thank Matthias and Robert for proof-reading this manuscript.

Furthermore, I thank Jürgen Saal for helping me to fill a gap in a proof of Chapter 5.

Moreover, I would like to thank my colleagues Eva Dintelmann, Bálint Farkas, Matthias Geißert, Karoline Götze, Robert Haller-Dintelmann, Tobias Hansel, Horst Heck, Okihiro Sawada, and Kyriakos Stavrakidis for the friendly atmosphere in our research group AG 4 in Darmstadt.

I also would like to thank the Graduiertenförderung der Technischen Universität Darmstadt for financial support and the Deutscher Akademischer Austauschdienst for the possibility to undertake two research visits at the Arizona State University.

Many thanks also go to my parents for their support and encouragement throughout my studies and my work on this thesis.

Finally, I thank my wife Nicole for her patience, her understanding, and her support during the entire period of my research.

Zusammenfassung

In dieser Arbeit befassen wir uns hauptsächlich mit der Stabilität geostrophischer Flüsse. Solche Flüsse sind zum Beispiel durch Strömungen in Ozeanen oder durch großräumige Luftbewegungen in der Atmosphäre gegeben.

Das grundlegende Modell für die mathematische Betrachtung geostrophischer Flüsse sind die Navier-Stokes-Gleichungen für viskose, inkompressible Fluide im dreidimensionalen Halbraum oder in einer Schicht. Dabei wird dieses Gleichungssystem um einen zusätzlichen Term erweitert, welcher die durch die Erdrotation verursachte Korioliskraft darstellt.

Für die entsprechenden Gleichungen im Halbraum existiert eine explizit gegebene stationäre Lösung, die sogenannte Ekman-Spirale. Wir beweisen in dieser Arbeit die Stabilität der Ekman-Spirale für den Fall, dass die Reynoldszahl des betrachteten hydrodynamischen Systems hinreichend klein ist. Hierzu untersuchen wir das Evolutionssystem, welches das Verhalten der gestörten Ekman-Spirale beschreibt. Wir zeigen, dass der zu diesem System assoziierte lineare Operator eine stark stabile Kontraktionshalbgruppe erzeugt. Dies ermöglicht es uns, schwache Lösungen dieses Systems zu konstruieren, die einen polynomiellen Abfall aufweisen. Hierzu beweisen wir ein Existenzresultat für schwache Lösungen von gestörten Navier-Stokes-Gleichungen.

Im Folgenden modifizieren wir die Ekman-Spirale, um eine stationäre Lösung der Navier-Stokes-Gleichungen mit Korioliskraft auf einer Schicht zu erhalten. Wieder betrachten wir das Evolutionssystem, welches das Verhalten einer Störung dieser Lösung bschreibt. Wie im Halbraumfall konstruieren wir schwache Lösungen dieses Systems. Aufgrund der Spektraleigenschaften des Laplace-Operators in Schichten weisen diese einen exponentiellen Abfall auf. Ausgehend von den Eigenschaften des Stokes-Operators in Schichten zeigen wir durch Störungstheorie, dass die zu diesem System assoziierte Halbgruppe globale L^p-L^q-Abschätzungen erfüllt. Dies ermöglicht es uns, eine starke Lösung des gestörten Systems zu konstruieren, die exponentiell abfällt.

Desweiteren untersuchen wir den Stokes-Operator in Gebieten mit kompaktem Rand. Wir beweisen, dass dieser im Fall eines hinreichend glatten Randes maximale Regularität hat. Dies ist für den Stokes-Operator in Halbräumen wohlbekannt. Durch eine Lokalisierung führen wir den Fall eines beschränkten Gebiets auf den Halbraumfall zurück. Da die übliche Lokalisierungstechnik die Divergenzfreiheit der Lösungen nicht erhält, modifizieren wir das Verfahren durch Anwendung des Bogovskiĭ-Operators. Im Fall eines Außenraumgebiets folgt die Behauptung aus dem Resultat im beschränkten Gebiet und bekannten Eigenschaften des Laplace-Operators im Ganzraum.

Chapter 1

Preliminaries

1.1 Notation

In this thesis the natural numbers are denoted by $\mathbb{N} := \{1, 2, \ldots\}$, whereas $\mathbb{N}_0 := \mathbb{N} \cup \{0\}$. Furthermore, Ω always denotes an open subset of the real vector space $\mathbb{R}^n$. For $m \in \mathbb{N}_0$, $C^m(\Omega)$ denotes the space of all m-times continuously differentiable functions and by $C_c^m(\Omega)$ we denote its subspace consisting of all functions in $C^m(\Omega)$ with compact support.

Let $m \in \mathbb{N}$. We say a domain Ω is *of class* C^m, if for every $x \in \partial\Omega$ there exist $r > 0$, a ball $B_r(x)$, and a real function g defined on a domain $\Omega' \subset \mathbb{R}^{n-1}$, such that $B_r(x) \cap \partial\Omega$ can be represented by an equation of the type $x_n = g(x_1, \ldots, x_{n-1})$, and such that each $y \in B_r(x) \cap \partial\Omega$ satisfies $x_n = g(y_1, \ldots, y_{n-1})$. We say Ω is a *Lipschitz domain*, if for every $x \in \partial\Omega$ we may choose a Lipschitz continuous function g in the foregoing construction.

For $1 \leq p < \infty$, $L^p(\Omega)$ denotes the Lebesgue space, which consists of all p-integrable functions and $L^\infty(\Omega)$ is the space of all functions u that satisfy $\|u\|_{L^\infty(\Omega)} := \operatorname{ess\,sup}_{x \in \Omega} |u(x)| < \infty$. Let $\mathcal{D}'(\Omega) := (C_c^m(\Omega))'$. We define $L^p_{\text{loc}}(\Omega) := \{u \in \mathcal{D}'(\Omega) : u \in L^p(K) \text{ for each compact } K \subset \Omega\}$.

Let A be a closed linear operator in a Banach space X. Then we denote by $D(A)$ the domain of A and by $\operatorname{rg}(A)$ its range.

In the following we introduce the notion of Sobolev spaces which are the function spaces we will work with during this thesis.

Definition 1.1. Let $\Omega \subset \mathbb{R}^n$ be a domain. Let $1 \leq p \leq \infty$ and $m \in \mathbb{N}_0$. Then the *Sobolev space of order* m is defined by

$$W^{m,p}(\Omega) := \{u \in L^p(\Omega) : D^\alpha u \in L^p(\Omega) \text{ for all } |\alpha| \leq m\},$$

where $D^\alpha u$ has to be understood in the sense of distributions. For $1 \leq p < \infty$ on $W^{m,p}(\Omega)$ we introduce the norm

$$\|u\|_{W^{m,p}(\Omega)} := \left(\sum_{|\alpha| \leq m} \|D^\alpha u\|_{L^p(\Omega)}^p \right)^{\frac{1}{p}}.$$

On $W^{m,\infty}(\Omega)$ the norm is given by

$$\|u\|_{W^{m,\infty}(\Omega)} := \sum_{|\alpha| \leq m} \|D^\alpha u\|_{L^\infty(\Omega)}.$$

Furthermore, $W_0^{m,p}(\Omega)$ denotes the closure of $C_c^\infty(\Omega)$ in $W^{m,p}(\Omega)$.

In the case $p = 2$, we will also write $W^{m,p}(\Omega) = H^m(\Omega)$ and $W_0^{m,p}(\Omega) = H_0^m(\Omega)$.

Moreover, the *Sobolev spaces of negative order* are defined as the dual spaces of $W_0^{m,p}(\Omega)$ by

$$W^{-m,p}(\Omega) := \left(W_0^{m,p'}(\Omega) \right)', \quad m \in \mathbb{N},\ 1 \leq p < \infty.$$

Note that $W^{m,p}(\Omega)$ equipped with the norm $\|\cdot\|_{W^{m,p}(\Omega)}$ becomes a Banach space. Thus, $W_0^{m,p}(\Omega)$ and $W^{-m,p}(\Omega)$ equipped with the respective norms are Banach spaces, too.

Remark 1.2. For sake of simplicity, we will frequently write $\|\cdot\|_{m,p}$ instead of $\|\cdot\|_{W^{m,p}(\Omega)}$ in the course of this thesis.

We will also make use of Sobolev spaces $W^{s,p}(\Omega)$ where $s \notin \mathbb{Z}$. We will introduce them using interpolation theory.

Let X and Y be Banach spaces, that form an interpolation couple. We denote the *real interpolation spaces* between X and Y by

$$(X, Y)_{\theta,p}, \quad \theta \in (0,1), \quad p \in [1, \infty].$$

Definition 1.3. Let $m \in \mathbb{N}$ and Ω be $\mathbb{R}^n$, $\mathbb{R}_+^n$, or a bounded domain of class C^m. Let $0 < s < m$, $s \notin \mathbb{N}$, and $\theta = \frac{s}{m}$. Then we define

$$W^{s,p}(\Omega) := \left(L^p(\Omega), W^{m,p}(\Omega)\right)_{\theta,p}.$$

$W_0^{s,p}(\Omega)$ denotes the closure of $C_c^\infty(\Omega)$ in $W^{s,p}(\Omega)$.

For $s < 0$, $s \notin \mathbb{Z}$, we define

$$W^{-s,p}(\Omega) := \left(W_0^{s,p'}(\Omega) \right)'.$$

Remark 1.4. Note that for $s \in \mathbb{N}$ the corresponding interpolation space would not coincide with the Sobolev spaces defined in Definition 1.1.

The *complex interpolation spaces* between X and Y will be denoted by

$$[X, Y]_\theta, \quad \theta \in (0, 1).$$

Remark 1.5. Let $m \in \mathbb{N}$, $s < m$ with $s \in \mathbb{N}$, and $\theta = \frac{s}{m}$. Then

$$[L^p(\Omega), W^{m,p}(\Omega)]_\theta = W^{s,p}(\Omega)$$

in the sense of Definition 1.1.

For a comprehensive examination of real and complex interpolation we refer to [Tri95].

In the course of this thesis we will encounter the question, whether $W^{s,p}(\Omega) = W_0^{s,p}(\Omega)$ or not. Therefore we state the following result, which gives an answer to this question in the halfspace case.

Proposition 1.6. *Let* $1 < p < \infty$ *and* $-1 + \frac{1}{p} < s < \frac{1}{p}$. *Then*

$$W^{s,p}(\mathbb{R}^n_+) = W_0^{s,p}(\mathbb{R}^n_+).$$

For a proof see [Tri95, Theorem 2.9.3].

For $1 \leq p < \infty$ and $m \in \mathbb{N}$, we shall further introduce the *homogeneous Sobolev spaces* $\widehat{W}^{m,p}(\Omega)$ consisting of all $L^1_{\text{loc}}(\Omega)$-functions u satisfying $\int_\Omega |\nabla^m u|^p \, dx < \infty$, where $\nabla^m u$ denotes the vector of all derivatives of u of order m. Then, $\widehat{W}^{m,p}(\Omega)$ modulo polynomials of order less than m becomes a Banach space when equipped with the norm

$$\|u\|_{\widehat{W}^{m,p}(\Omega)} := \left(\int_\Omega |\nabla^m u|^p \, dx \right)^{\frac{1}{p}}.$$

Note that, if Ω is a bounded domain, $\widehat{W}^{1,p}(\Omega)$ is isomorphic to $W^{1,p}(\Omega) \cap L^p_0(\Omega)$, where

$$L^p_0(\Omega) := \{u \in L^p(\Omega) : \int_\Omega u \, dx = 0\}.$$

On the other hand, $W^{1,p}(\Omega)$ can be embedded continuously into the homogeneous space $\widehat{W}^{1,p}(\Omega)$ for arbitrary domains Ω.

$\widehat{W}^{m,p}(\Omega)$ contains $C_c^\infty(\overline{\Omega})$ as a dense subspace for $1 \leq p < \infty$.

1.2 Operator theory and semigroups

We start this section with a survey of the concept of operator semigroups, which play an important role in the course of this thesis. In a Banach space X, consider the abstract Cauchy problem

$$\begin{cases} u'(t) &= Au(t), \quad t > 0, \\ u(0) &= u_0, \end{cases} \tag{1.1}$$

where A is a closed linear operator in X. The theory of operator semigroups yields conditions for (unique) solvability of (1.1) in terms of the spectral properties of A.

Definition 1.7. Let X be a Banach space. A family $T = \{T(t) : t \geq 0\}$ of bounded linear operators on X is called a *strongly continuous semigroup* (or C_0*-semigroup*), if it satisfies the following properties.

1. $T(0) = \mathrm{Id}$.
2. $T(s+t) = T(s)T(t)$ for all $s, t \geq 0$.
3. The map $t \mapsto T(t)x$ is continuous for all $x \in X$.

For every C_0-semigroup T there exist constants $M \geq 1$ and $\omega \in \mathbb{R}$, such that

$$\|T(t)\| \leq Me^{\omega t}, \quad t \geq 0.$$

The infimum over all possible ω is called the *growth bound* $\omega(T)$ of the semigroup. A semigroup T is called

1. *quasicontractive*, if we can choose $M = 1$, and
2. *contractive*, if $\|T(t)\| \leq 1$ for all $t \geq 0$.

Definition 1.8. Let X be a Banach space and let T be a C_0-semigroup on X. The operator $A : D(A) \to X$ defined by

$$Ax := \lim_{h \to 0} \frac{1}{h} (T(h)x - x)$$

with

$$D(A) := \{x \in X : \lim_{h \to 0} \frac{1}{h} (T(h)x - x) \text{ exists}\}$$

is called the *generator* of T.

Remark 1.9. 1. The generator of a C_0-semigroup is closed, densely defined, and unique.

2. Let A be the generator of the C_0-semigroup T. Then $(\omega(T), \infty) \subset \rho(A)$ and the resolvent of A can be obtained via the Laplace transform by

$$R(\lambda, A) := (\lambda - A)^{-1} = \int_0^\infty e^{-\lambda t} T(t)\, \mathrm{d}t, \quad \operatorname{Re}\lambda > \omega(T).$$

3. Let A be the generator of a C_0-semigroup and $u_0 \in D(A)$. Then $u(t) = T(t)u_0$ is the unique solution to (1.1) satisfying $u \in C^1(\mathbb{R}_+, X)$ and $u(t) \in D(A)$ for $t > 0$.

For details see [ABHN01] or [EN00].

There are different methods to determine, whether an operator is the generator of a C_0-semigroup. In the following we will state a result, which gives an answer in the contractive case. We proceed with the following definition.

Definition 1.10. A linear operator A in X is called *dissipative*, if

$$\|(\lambda - A)x\| \geq \lambda \|x\|$$

for all $\lambda > 0$ and $x \in D(A)$.

In Hilbert spaces there is an equivalent characterization of dissipative operators, which is easier to check.

Proposition 1.11. *An operator A in a Hilbert space H is dissipative, if and only if for every $x \in D(A)$ we have*

$$\operatorname{Re}\langle Ax, x\rangle \leq 0.$$

For a proof, see [ABHN01, Lemma 3.4.2].

A useful tool in semigroup theory is the following generation theorem of Lumer and Phillips.

Proposition 1.12. *For a densely defined, dissipative operator A the following assertions are equivalent:*

1. *The closure $\overline{A}$ of A is the generator of a contractive semigroup.*
2. $\operatorname{rg}(\lambda - A)$ *is dense for one (hence all) $\lambda > 0$.*

For a proof, we refer to [ABHN01, Theorem 3.4.5].

In the next step we would like to extend the considered families of operators to complex parameters.

Definition 1.13. Let $\theta \in (0, \frac{\pi}{2}]$. A semigroup T is called *analytic of angle* θ, if it has an analytic extension to the sector

$$\Sigma_\theta := \{z \in \mathbb{C} \setminus \{0\} : |\arg z| < \theta\},$$

which is bounded on $\Sigma_{\theta'} \cap \{z \in \mathbb{C} : |z| \leq 1\}$ for all $0 < \theta' < \theta$.

If T has a bounded analytic extension to $\Sigma_{\theta'}$ for each $\theta' \in (0, \theta)$, we call T a *bounded analytic semigroup of angle* θ.

The following result gives a characterization of generators of analytic semigroups in terms of resolvent estimates.

Proposition 1.14. *Let $\theta \in (0, \frac{\pi}{2}]$. For a closed linear operator A in a Banach space X the following assertions are equivalent.*

1. *A generates a bounded analytic semigroup of angle θ on X.*
2. *$\Sigma_{\theta+\frac{\pi}{2}} \subset \rho(A)$ and for all $0 < \theta' < \theta$ there exists a constant M, such that*

$$\|R(\lambda, A)\| \leq \frac{M}{|\lambda|}$$

for all $\lambda \in \Sigma_{\theta'+\frac{\pi}{2}}$.

A proof can be found for example in [ABHN01, Theorem 3.7.11].

Remark 1.15.

1. If A is the generator of a bounded strongly continuous semigroup T on a Banach space X, then T extends to a bounded analytic semigroup, if and only if

$$\sup_{t>0} \|tAT(t)\| < \infty.$$

2. For every analytic (not necessarily bounded) semigroup T there is a constant $C > 0$, such that

$$\|tAT(t)\| < C, \qquad 0 < t \leq 1.$$

3. An operator A is the generator of an analytic semigroup, if and only if there exists $c \geq 0$, such that $A - c$ is the generator of a bounded analytic semigroup.
4. If T is an analytic semigroup, then $\operatorname{rg}(T(t)) \subset D(A)$ for all $t > 0$.
5. Let A be a closed, densely defined operator in X and suppose that the adjoint operator A' is densely defined in X'. Then A generates an analytic semigroup on X, if and only if A' is the generator of an analytic semigroup on X'.

For details we refer to [ABHN01] or [EN00].

Definition 1.16. A closed operator A in a Banach space X is called *sectorial*, if it satisfies the following conditions:

1. A is densely defined and injective;
2. $(-\infty, 0) \subset \rho(A)$ and there is some $M > 0$, such that

$$\|\lambda(\lambda + A)^{-1}\| \leq M, \quad \lambda > 0.$$

Remark 1.17. If A is a sectorial operator, then $\mathrm{rg}(A)$ is dense.

For a sectorial operator A there exists $\phi \in [0, \pi)$, such that the sector $\Sigma_{\pi-\phi}$ is contained in $\rho(-A)$, and $\sup\{\|\lambda(\lambda + A)^{-1}\| : \lambda \in \Sigma_{\pi-\phi}\} < \infty$. The infimum over all possible angles ϕ is called *spectral angle* of A and is denoted by ϕ_A. Then we have that $\sigma(A) \setminus \{0\} \subset \Sigma_{\phi_A}$. The class of all sectorial operators on X will be denoted by $\mathcal{S}(X)$.

Remark 1.18. An operator A is sectorial of angle $\phi_A < \frac{\pi}{2}$, if and only if $-A$ is the generator of a bounded analytic semigroup.

The next result deals with perturbations of sectorial operators.

Proposition 1.19. *Suppose $A \in \mathcal{S}(X)$ and B is a closed, linear operator satisfying $D(A) \subset D(B)$ and $\|Bx\| \leq b\|Ax\|$ for all $x \in D(A)$ with some constant $b \geq 0$. Furthermore, let*

$$C_{\pi-\phi}(A) = \sup_{\lambda \in \Sigma_{\pi-\phi}} \|A(\lambda + A)^{-1}\|.$$

Then $bC_0(A) < 1$ implies $A + B \in \mathcal{S}(X)$, and the spectral angle ϕ_{A+B} of $A + B$ satisfies

$$\phi_{A+B} \leq \inf\{\phi > \phi_A : bC_{\pi-\phi}(A) < 1\}.$$

For a proof we refer to [DHP03, Theorem 1.5].

In the following we introduce a functional calculus for sectorial operators. For this purpose, for $\phi \in (0, \pi]$ we define the algebra

$$\mathcal{H}^\infty(\Sigma_\phi) := \{f : \Sigma_\phi \to \mathbb{C} : f \text{ is bounded and analytic}\}$$

and the subalgebra

$$\mathcal{H}_0^\infty(\Sigma_\phi) := \left\{f \in \mathcal{H}^\infty(\Sigma_\phi) : |f(z)| \leq C\frac{|z|^s}{1 + |z|^{2s}} \text{ for some } C \geq 0, s > 0\right\}.$$

Consider a sectorial operator A with spectral angle ϕ_A, $\phi > \delta > \phi_A$, and set $\Gamma := \{\gamma(t) : t \in \mathbb{R}\}$ with

$$\gamma(t) = \begin{cases} te^{-i\delta}, & t \geq 0, \\ -te^{i\delta}, & t < 0. \end{cases}$$

Note that $\Gamma \setminus \{0\} \subset \rho(A)$. Due to the resolvent estimate satisfied by A, for $f \in \mathcal{H}_0^\infty(\Sigma_\phi)$, we may define $f(A)$ via the Dunford integral by

$$f(A) := \frac{1}{2\pi i} \int_\Gamma f(\lambda) R(\lambda, A) \, \mathrm{d}\lambda.$$

Definition 1.20. A sectorial operator A in a Banach space X is said to *admit a bounded $\mathcal{H}^\infty$-calculus*, if there exist an angle $\phi \in (\phi_A, \pi)$ and a constant $M_\phi > 0$, such that

$$\|f(A)\| \leq M_\phi \|f\|_\infty \tag{1.2}$$

for all $f \in \mathcal{H}_0^\infty(\Sigma_\phi)$. The infimum over all angles, for which the estimate (1.2) is valid, is called the *$\mathcal{H}^\infty$-angle* and will be denoted by ϕ_A^∞. We denote the set of all sectorial operators admitting a bounded $\mathcal{H}^\infty$-calculus on X by $\mathcal{H}^\infty(X)$.

Remark 1.21. If $A \in \mathcal{H}^\infty(X)$, then the functional calculus for A extends uniquely to $\mathcal{H}^\infty(\Sigma_\phi)$ and (1.2) holds for all $f \in \mathcal{H}^\infty(\Sigma_\phi)$ (see [DHP03]).

The following proposition gives an example of linear operators, that admit a bounded $\mathcal{H}^\infty$-calculus (see for example [DHP03]).

Proposition 1.22. *Let A be a closed linear operator in a Hilbert space H, such that $-A$ is dissipative. Then $A \in \mathcal{H}^\infty(H)$.*

Now, we focus our interest on functions of the form $f(z) = z^{is}$ with $s \in \mathbb{R}$, which are bounded and holomorphic on each sector Σ_ϕ for $0 < \phi < \pi$. By the method described above we are able to define A^{is} for sectorial operators. These operators will be unbounded in general. We will introduce a further subclass of sectorial operators, for which A^{is} is bounded.

Definition 1.23. Let A be a sectorial operator in a Banach space X. We say that A has *bounded imaginary powers*, if $A^{is} \in \mathscr{L}(X)$ for all $s \in \mathbb{R}$ and there exists $C > 0$, such that for some $\varepsilon > 0$

$$\|A^{is}\| \leq C, \quad s \in [-\varepsilon, \varepsilon].$$

The set of all operators having bounded imaginary powers on X will be denoted by $\mathrm{BIP}(X)$.

For a sectorial operator A the family $(A^{is})_{s\in\mathbb{R}}$ is a strongly continuous group of bounded linear operators on X, if and only if $A \in \mathrm{BIP}(X)$ (see [DHP03, Proposition 2.2]). In this case its growth bound

$$\theta_A := \limsup_{s\in\mathbb{R}} \frac{1}{|s|} \log \|A^{is}\|$$

is called the *power angle* of A.

A useful application of the class $\mathrm{BIP}(X)$ concerns the representation of the domain of fractional powers of a sectorial operator in terms of complex interpolation spaces.

Proposition 1.24. *Let $A \in \mathrm{BIP}(X)$ and $\alpha \in (0,1)$. Then*

$$D(A^\alpha) = [X, D(A)]_\alpha.$$

For a proof we refer to [Tri95, Theorem 1.15.3].

Furthermore, we will give a sufficient condition on an operator A to admit bounded imaginary powers.

Proposition 1.25. *Let A be a positive self-adjoint operator in a Hilbert space H. Then $A \in \mathrm{BIP}(H)$ with $\theta_A = 0$ and $\|A^{is}\| \leq 1$ for all $s \in \mathbb{R}$.*

For a proof see [Ama95, Example III.4.7.3].

Next we introduce the notion of $\mathcal{R}$-bounded families of operators.

Definition 1.26. Let X and Y be Banach spaces. A family of operators $\mathcal{T} \subset \mathscr{L}(X,Y)$ is called *$\mathcal{R}$-bounded*, if there is a constant $C > 0$ and some $p \in [1,\infty)$, such that for each $N \in \mathbb{N}$, every choice $T_j \in \mathcal{T}$, $x_j \in X$ with $j = 1,\dots,N$, and for all independent, symmetric, $\{-1,1\}$-valued random variables ε_j, $j = 1,\dots,N$, on a probability space $(\mathcal{M},\mathcal{A},\mu)$ the inequality

$$\left\| \sum_{j=1}^{N} \varepsilon_j T_j x_j \right\|_{L^p(\Omega,Y)} \leq C \left\| \sum_{j=1}^{N} \varepsilon_j x_j \right\|_{L^p(\Omega,X)}$$

is valid. The infimum of all possible C is called *$\mathcal{R}$-bound* of $\mathcal{T}$ and is denoted by $\mathcal{R}(\mathcal{T})$.

For details and calculation rules concerning $\mathcal{R}$-bounds we refer to [DHP03].

In the course of this thesis we will have to link aspects of interpolation theory with the concept of $\mathcal{R}$-boundedness. In this context we will make use of the following result (cf. [GHH+08]).

Lemma 1.27. *Let X, Y, Z be Banach spaces, $\theta \in (0,1)$, and $p \in (1,\infty)$. We assume that Y and Z form an interpolation couple. Furthermore, let $\mathcal{F} \subset \mathscr{L}(X,Y) \cap \mathscr{L}(X,Z)$ be an $\mathcal{R}$-bounded set of operators with $\mathcal{R}$-bounds $\mathcal{R}_{X\to Y}(\mathcal{F})$ and $\mathcal{R}_{X\to Z}(\mathcal{F})$, respectively. Then $\mathcal{F} \subset \mathscr{L}(X,(Y,Z)_{\theta,p})$ is $\mathcal{R}$-bounded and there exists $C > 0$, such that*

$$\mathcal{R}_{X\to(Y,Z)_{\theta,p}}(\mathcal{F}) \leq C\mathcal{R}_{X\to Y}(\mathcal{F})^{1-\theta}\mathcal{R}_{X\to Z}(\mathcal{F})^{\theta}. \tag{1.3}$$

The same result holds, if we replace the real interpolation space $(Y,Z)_{\theta,p}$ by the complex interpolation space $[Y,Z]_\theta$, $\theta \in (0,1)$.

Proof. Let $N \in \mathbb{N}$. For $j = 1,\dots,N$ let $T_j \in \mathcal{F}$, $f_j \in X$, and ε_j independent, symmetric $\{-1,1\}$-valued random variables on a probability space $(\mathcal{M},\mathcal{A},\mu)$. Using an interpolation argument (see [Tri95, Theorem 1.3.3(g)]), we obtain

$$\Big\|\sum_{j=1}^N \varepsilon_j(\omega)T_jf_j\Big\|_{(Y,Z)_{\theta,p}} \leq C\Big\|\sum_{j=1}^N \varepsilon_j(\omega)T_jf_j\Big\|_Y^{1-\theta}\Big\|\sum_{j=1}^N \varepsilon_j(\omega)T_jf_j\Big\|_Z^{\theta}. \tag{1.4}$$

Now, it follows from Hölder's inequality that

$$\begin{aligned}
&\int_{\mathcal{M}}\Big\|\sum_{j=1}^N \varepsilon_j(\omega)T_jf_j\Big\|_Y^{1-\theta}\Big\|\sum_{j=1}^N \varepsilon_j(\omega)T_jf_j\Big\|_Z^{\theta}\,\mathrm{d}\omega \\
&\leq\; C\Big(\int_{\mathcal{M}}\Big\|\sum_{j=1}^N \varepsilon_j(\omega)T_jf_j\Big\|_Y\,\mathrm{d}\omega\Big)^{1-\theta}\Big(\int_{\mathcal{M}}\Big\|\sum_{j=1}^N \varepsilon_j(\omega)T_jf_j\Big\|_Z\,\mathrm{d}\omega\Big)^{\theta}
\end{aligned}$$

holds. Combining this inequality with (1.4), we obtain by the $\mathcal{R}$-boundedness of $\mathcal{F}$ in $\mathscr{L}(X,Y)$ and $\mathscr{L}(X,Z)$

$$\begin{aligned}
&\int_{\mathcal{M}}\Big\|\sum_{j=1}^N \varepsilon_j(\omega)T_jf_j\Big\|_{(Y,Z)_{\theta,p}}\,\mathrm{d}\omega \\
&\leq\; C\mathcal{R}_{X\to Y}(\mathcal{F})^{1-\theta}\Big(\int_{\mathcal{M}}\Big\|\sum_{j=1}^N \varepsilon_j(\omega)f_j\Big\|_X\,\mathrm{d}\omega\Big)^{1-\theta} \\
&\qquad\cdot\mathcal{R}_{X\to Z}(\mathcal{F})^{\theta}\Big(\int_{\mathcal{M}}\Big\|\sum_{j=1}^N \varepsilon_j(\omega)f_j\Big\|_X\,\mathrm{d}\omega\Big)^{\theta} \\
&=\; C\mathcal{R}_{X\to Y}(\mathcal{F})^{1-\theta}\mathcal{R}_{X\to Z}(\mathcal{F})^{\theta}\Big(\int_{\mathcal{M}}\Big\|\sum_{j=1}^N \varepsilon_j(\omega)f_j\Big\|_X\,\mathrm{d}\omega\Big).
\end{aligned}$$

If we consider $\mathcal{R}_{X\to[Y,Z]_\theta}(\mathcal{F})$, the claim follows in the same way using the corresponding interpolation argument for complex interpolation spaces (see [Tri95, Theorem 1.9.3(f)]). □

The concept of $\mathcal{R}$-boundedness allows us to introduce two further classes of operators, that are closely related to the ones already introduced in this section.

Definition 1.28. Let A be a sectorial operator in a Banach space X.

1. Let
$$\mathcal{R}_A(\theta) := \mathcal{R}\{\lambda(\lambda + A)^{-1} : \lambda \in \mathbb{C} \setminus \{0\}, |\arg \lambda| \leq \theta\}.$$
Then A is called *$\mathcal{R}$-sectorial*, if
$$\mathcal{R}_A(0) < \infty.$$
The *$\mathcal{R}$-angle* $\phi_A^{\mathcal{R}}$ of A is defined by
$$\phi_A^{\mathcal{R}} := \inf\{\theta \in (0, \pi) : \mathcal{R}_A(\pi - \theta) < \infty\}.$$
The class of all $\mathcal{R}$-sectorial operators on X will be denoted by $\mathcal{RS}(X)$.

2. Let $A \in \mathcal{H}^\infty(X)$. The operator A is said to admit an *$\mathcal{R}$-bounded $\mathcal{H}^\infty$-calculus*, if the set
$$\{h(A) : h \in \mathcal{H}^\infty(\Sigma_\theta), \|h\|_\infty \leq 1\}$$
is $\mathcal{R}$-bounded for some $\theta > 0$. The *$\mathcal{RH}^\infty$-angle* $\phi_A^{\mathcal{R}\infty}$ of A is defined as the infimum of such angles θ. The class of all operators admitting an $\mathcal{R}$-bounded $\mathcal{H}^\infty$-calculus in X will be denoted by $\mathcal{RH}^\infty(X)$.

Remark 1.29. For the classes of operators introduced so far the inclusions
$$\mathcal{RH}^\infty(X) \subset \mathcal{H}^\infty(X) \subset \mathrm{BIP}(X) \subset \mathcal{RS}(X) \subset \mathcal{S}(X)$$
and the inequalities
$$\phi_A^{\mathcal{R}\infty} \geq \phi_A^\infty \geq \theta_A \geq \phi_A^{\mathcal{R}} \geq \phi_A$$
hold.

Next, we state a result that concerns perturbations of $\mathcal{R}$-sectorial operators.

Proposition 1.30. *Suppose A is an $\mathcal{R}$-sectorial operator in a Banach space X, and let B be closed and linear, such that*
$$D(A) \subset D(B), \quad \|Bx\| \leq \alpha\|Ax\| + \beta\|x\|, \quad x \in D(A),$$
with some constants $\alpha, \beta > 0$. Furthermore, for some $\theta \in (0, \pi)$ let
$$a := \mathcal{R}\{\lambda(\lambda + A)^{-1} : \lambda \in \Sigma_\theta\}.$$

Then

$$\mathcal{R}\{\lambda(\lambda+\mu+A+B)^{-1} : \lambda \in \Sigma_\theta\} < \infty,$$

whenever $\alpha < \frac{1}{(1+a)C_A}$ *and* $\mu > \beta M_A \frac{1+a}{1-\alpha C_A(1+a)}$, *where*

$$C_A = \sup_{r>0} \|A(r+A)^{-1}\|$$

and

$$M_A = \sup_{r>0} \|r(r+A)^{-1}\|.$$

A proof can be found in [DHP03, Prop. 4.3].

Now, will we describe the relation between $\mathcal{R}$-boundedness and maximal L^p-regularity. Consider the abstract Cauchy problem

$$\begin{cases} u'(t) + Au(t) &= f(t), \quad t > 0, \\ u(0) &= 0, \end{cases} \tag{1.5}$$

where A denotes a sectorial operator in a Banach space X of angle $\phi_A < \frac{\pi}{2}$. The solution of this problem for given $f \in L^p(\mathbb{R}_+, X)$, $1 < p < \infty$, can be represented by the variation of constant formula

$$u(t) = \int_0^t e^{-As} f(t-s)\, ds, \quad t \geq 0.$$

We say that (1.5) has *maximal regularity of type* L^p, if $Au \in L^p(\mathbb{R}_+, X)$, whenever $f \in L^p(\mathbb{R}_+, X)$. This means that there is a constant $C > 0$, such that

$$\|u'\|_{L^p(\mathbb{R}_+,X)} + \|Au\|_{L^p(\mathbb{R}_+,X)} \leq C\|f\|_{L^p(\mathbb{R}_+,X)}.$$

Moreover, if the abstract Cauchy problem (1.5) has maximal regularity of type L^p, then it admits a unique solution $u \in L^p(\mathbb{R}_+, D(A)) \cap W^{1,p}(\mathbb{R}_+, X)$ (for details we refer to [DHP03]).

By classical results (see [Sob64]) this property is independent of p. Thus, in the following we will write *maximal regularity* instead of maximal regularity of type L^p. Furthermore, we say that an operator A has maximal regularity, if the associated abstract Cauchy problem has maximal regularity.

The relation between $\mathcal{R}$-boundedness and the property of maximal regularity was first observed by Weis in [Wei01]. He could give a characterization of maximal regularity of an operator A in terms of $\mathcal{R}$-boundedness of the set $\{\lambda(i\lambda - A)^{-1} : \lambda \in \mathbb{R}\}$ in case the underlying Banach space is a UMD space. For details on UMD spaces we refer to [Ama95, III.4.5]. In particular, for $1 < p < \infty$ and $\Omega \subset \mathbb{R}^n$ the spaces $L^p(\Omega)$ and $L^p_\sigma(\Omega)$ are UMD spaces. Precisely, we have the following result.

Proposition 1.31. *Let A be a sectorial operator of angle $\phi_A < \frac{\pi}{2}$ in an UMD space X, $1 < p < \infty$. Then the following statements are equivalent:*

1. *A has maximal regularity.*
2. *A is $\mathcal{R}$-sectorial.*
3. *The set $\{\lambda(i\lambda - A)^{-1} : \lambda \in \mathbb{R}\}$ is $\mathcal{R}$-bounded.*

For a proof we refer to [Wei01].

In many cases an estimate as in Proposition 1.31 can only be shown for the shifted operator $A + r$ for some $r > 0$. In [Dor93] it is shown that in this case the property of maximal regularity is preserved, if we restrict (1.5) to an arbitrary finite time interval $(0, T)$, $T \in \mathbb{R}_+$. Furthermore, a result is given that allows to obtain maximal regularity for $t \in \mathbb{R}_+$ from maximal regularity on a finite interval.

Proposition 1.32. *Let X be a complex Banach space and A a linear, closed, and densely defined operator on X. If A admits maximal regularity on $(0, T)$ for all $T \in \mathbb{R}_+$ and the growth bound of the semigroup generated by $-A$ is negative, then A admits maximal regularity on $\mathbb{R}_+$.*

A proof can be found in [Dor93, Thm. 2.4].

1.3 Helmholtz decomposition

In the study of problems in the context of the Stokes and Navier-Stokes equations, the notion of solenoidal vector fields plays a natural role. Of special interest is the question, if for an open set $\Omega \subset \mathbb{R}^n$ a given vector field $u \in L^p(\Omega)^n$ may be decomposed in a solenoidal part v and a potential part $w = \nabla p$ for some scalar function $p \in \widehat{W}^{1,p}(\Omega)$. Such a decomposition will be essential for the later definition of the so-called Stokes operator.

For $1 < p < \infty$ we define the space $L^p_\sigma(\Omega)$ by

$$L^p_\sigma(\Omega) := \overline{C^\infty_{c,\sigma}(\Omega)}^{\|\cdot\|_{L^p(\Omega)^n}}. \tag{1.6}$$

Here $C^\infty_{c,\sigma}(\Omega)$ denotes the set of all vector fields $f \in C^\infty_c(\Omega)^n$, such that $\operatorname{div} f = 0$. We have that $L^p_\sigma(\Omega)$ is a closed subspace of $L^p(\Omega)^n$ for every $1 < p < \infty$. Hence, $L^2_\sigma(\Omega)$ is even a Hilbert space. Defining a complementary space $G_2(\Omega) := L^2_\sigma(\Omega)^\perp$ we obtain a decomposition

$$L^2(\Omega)^n = L^2_\sigma(\Omega) \oplus G_2(\Omega), \tag{1.7}$$

where $\oplus$ denotes the direct sum operation. This decomposition is called the *Helmholtz decomposition.* Associated to this decomposition there is a linear, bounded, and orthogonal projection $P_2 : L^2(\Omega)^n \to L^2_\sigma(\Omega)$, the so-called *Helmholtz projection.* In the Hilbert space case this decomposition works independently of the properties of the underlying domain.

A direct generalization of this construction to the $L^p(\Omega)$-setting fails, since in $L^p(\Omega)^n$ not every closed subspace is automatically complementary for $p \neq 2$. Nevertheless, for a large class of domains the existence of the Helmholtz projection in $L^p(\Omega)^n$ is known. In particular, we have the following result.

Proposition 1.33. *For $n \geq 2$ let $\Omega \subset \mathbb{R}^n$ be the whole space, the half-space, an infinite layer, or a domain of class C^2 with compact boundary. Then for every $p \in (1, \infty)$ the Helmholtz decomposition holds in $L^p(\Omega)^n$.*

For the cases of the whole space, the halfspace, and domains of class C^2, a proof of this result is given in [Gal94a]. For the problem considered in infinite layers we refer to [Miy94] and [Far03].

In the following we collect some helpful properties of the Helmholtz projection (for details see [Gal94a]).

Proposition 1.34. *Let $\Omega \subset \mathbb{R}^n$ be one of the above domains. Then the following statements hold:*

1. *For the annihilators of the decomposing spaces we obtain $L^p_\sigma(\Omega)^\perp = G_{p'}(\Omega)$ and $G_p(\Omega)^\perp = L^{p'}_\sigma(\Omega)$.*
2. *$(P_{\Omega,p})_{p\in(1,\infty)}$, where $P_{\Omega,p} : L^p(\Omega)^n \to L^p_\sigma(\Omega)$ is the associated Helmholtz decomposition, is a compatible family of linear and bounded operators, i.e. for $f \in L^p(\Omega)^n \cap L^q(\Omega)^n$ we have $P_{\Omega,p}f = P_{\Omega,q}f$ for $p, q \in (1, \infty)$.*
3. *We have $P'_{\Omega,p} = P_{\Omega,p'}$, $L^p_\sigma(\Omega)' = L^{p'}_\sigma(\Omega)$, and $G_p(\Omega)' = G_{p'}(\Omega)$.*

Remark 1.35. The first assertion of Proposition 1.34 implies that for $1 < p < \infty$

$$L^p_\sigma(\Omega) = \{u \in L^p(\Omega)^n : \langle u, \nabla p\rangle = 0 \text{ for all } p \in \widehat{W}^{1,p'}(\Omega)\}.$$

In the course of this thesis the Stokes operator with Dirichlet boundary conditions in $L^p_\sigma(\Omega)$ will appear frequently. Therefore, we will give a precise definition of this operator.

Definition 1.36. Let $\Omega \subset \mathbb{R}^n$, $n \geq 2$, be the whole space $\mathbb{R}^n$, the halfspace, an infinite layer, or a bounded or exterior domain of class C^2. Let $1 < p < \infty$. Then the *Stokes operator* with homogeneous Dirichlet boundary conditions is defined by

$$\left\{\begin{array}{rcl} A_{\Omega,p}u & := & -P_{\Omega,p}\Delta u, \\ D(A_{\Omega,p}) & := & W^{2,p}(\Omega)^n \cap W^{1,p}_0(\Omega)^n \cap L^p_\sigma(\Omega). \end{array}\right.$$

Remark 1.37. For convenience, from now on we will simply use the notation $P = P_{\Omega,p}$. If there is no danger of confusion, we will write A instead of $A_{\Omega,p}$.

1.4 The Bogovskiĭ operator

For the purposes of this thesis we will frequently make use of the so-called Bogovskiĭ operator. This operator acts as a right inverse of the divergence operator in certain function spaces. Basically, we are interested in solutions of the divergence equation

$$\left\{ \begin{array}{rcll} \operatorname{div} u &=& g & \text{in } \Omega, \\ u &=& 0 & \text{on } \partial\Omega, \end{array} \right. \tag{1.8}$$

where $\Omega \subset \mathbb{R}^n$, $n \geq 2$, is a domain and g a given scalar valued function.

Bogovskiĭ proved the existence and a-priori estimates of solutions to (1.8) in the scale of Sobolev spaces of positive order using an explicit representation formula for u on bounded, star shaped domains. His method even applies to bounded Lipschitz domains provided $g \in L^p_0(\Omega)$, $1 < p < \infty$. Furthermore, a solution operator B_Ω of (1.8) can be constructed (for details see [Bog79], [Gal94a]). Geissert, Heck, and Hieber proved in [GHH06b] that this solution operator B_Ω can be extended continuously to an operator mapping from $W^{s,p}_0(\Omega)$ to $W^{s+1,p}_0(\Omega)^n$ where $s > -2 + \frac{1}{p}$. Their approach is based on a detailed examination of the adjoint kernel of K, where K denotes the kernel associated to B_Ω, see also [BS90]. Summing up we have the following result.

Proposition 1.38. *Let $\Omega \subset \mathbb{R}^n$ be a bounded Lipschitz domain. Then there exists a mapping $B_\Omega : C^\infty_c(\Omega) \to C^\infty_c(\Omega)^n$, such that for $g \in C^\infty_c(\Omega) \cap L^p_0(\Omega)$*

$$\operatorname{div} B_\Omega g = g$$

holds. Moreover, B_Ω can be extended continuously to a bounded operator from $W^{s,p}_0(\Omega)$ to $W^{s+1,p}_0(\Omega)^n$, such that

$$\|B_\Omega g\|_{W^{s+1,p}_0(\Omega)^n} \leq C \|g\|_{W^{s,p}_0(\Omega)}, \quad g \in W^{s,p}_0(\Omega), \tag{1.9}$$

provided $s > -2 + \frac{1}{p}$.

Note that no compatibility condition on g is needed in the above estimate (1.9). However, if $\int_\Omega g \, \mathrm{d}x \neq 0$, then $B_\Omega g$ is not a solution to the divergence problem (1.8).

1.5 The Gagliardo-Nirenberg inequality

In the following section we will give a Sobolev type inequality, which holds on so-called (ε, ∞) domains and will be important in the course of this thesis.

To motivate this section, let $p < n$, $\frac{1}{p^*} = \frac{1}{p} - \frac{1}{n}$, and $\Omega \subset \mathbb{R}^n$. It is well known that in this case the classical Sobolev inequality

$$\|u\|_{L^{p^*}(\Omega)} \leq C\|\nabla u\|_{L^p(\Omega)} \tag{1.10}$$

holds for all $u \in C_c^\infty(\Omega)$. Thus, it holds for all $u \in W_0^{1,p}(\Omega) = \overline{C_c^\infty(\Omega)}^{\|\cdot\|_{1,p}}$, too. Since $W_0^{1,p}(\Omega) \neq W^{1,p}(\Omega)$ in general, the question arises whether (1.10) is valid in a domain Ω for functions lying merely in $W^{1,p}(\Omega)$. In this context the notion of unbounded (ε, ∞) domains comes into play.

Definition 1.39. Let $\Omega \subset \mathbb{R}^n$. For a rectifiable arc $\Gamma \subset \mathbb{R}^n$ let $L(\Gamma)$ denote its Euclidian arclength. For $x \in \Omega$ let $d(x) := \inf_{y\in\Omega^c} |x - y|$ denote the Euclidian distance from x to $\partial\Omega$. We say that Ω is an (ε, ∞) domain, if there is an $\varepsilon > 0$ having the following property:
For all $x, y \in \Omega$ there is a rectifiable arc $\Gamma \subset \Omega$ joining x to y and satisfying

$$L(\Gamma) \leq \frac{1}{\varepsilon}|x - y|,$$

as well as

$$d(z) \geq \varepsilon\frac{|x - z||y - z|}{|x - y|}, \quad z \in \Gamma.$$

For unbounded (ε, ∞) domains we can obtain much more general inequalities than (1.10). In fact, the following Gagliardo-Nirenberg inequality holds.

Proposition 1.40. *Let $1 \leq r, q \leq \infty$ and $\Omega \subset \mathbb{R}^n$ be an (ε, ∞) domain. Furthermore, let $j, m \in \mathbb{N}_0$ with $0 \leq j < m$. If $1 \leq p \leq \infty$ and $\frac{j}{m} \leq a < 1$, such that*

$$\frac{1}{p} = \frac{j}{n} + a\left(\frac{1}{r} - \frac{m}{n}\right) + (1 - a)\frac{1}{q},$$

then

$$\|\nabla^j u\|_p \leq C\|\nabla^m u\|_r^a \|u\|_q^{1-a} \tag{1.11}$$

holds for all $u \in \widehat{W}^{m,r}(\Omega) \cap L^q(\Omega)$.
With the only exception that $\frac{m-j-n}{r} \in \mathbb{N}_0$, (1.11) holds true for all $a \in [\frac{j}{m}, 1]$.

Proof. From [Fri69, Theorem 9.3] we obtain the validity of (1.11) for $\Omega = \mathbb{R}^n$ and all $u \in C_c^\infty(\mathbb{R}^n)$. Therefore (1.11) holds for all $u \in \widehat{W}^{m,r}(\Omega) \cap L^q(\Omega)$, too. By the [Chu92, Theorem 1.9] the assertion follows. □

In the following we give some examples of unbounded (ε, ∞) domains, which will be of interest in this thesis.

Example 1.41.

1. Consider $\Omega = \mathbb{R}^n$. Since $\partial\Omega = \emptyset$, we have $d(x) = \infty$ for all $x \in \Omega$. Hence, $\mathbb{R}^n$ is obviously an (ε, ∞) domain.

2. In the case $\Omega = \mathbb{R}^n_+ := \{x \in \mathbb{R}^n : x_n > 0\}$ consider the following construction:
   For any $x \in \mathbb{R}^n_+$, let $\bar{x} := (x_1, \ldots, x_{n-1}, 0)$ denote the orthogonal projection onto the plane $\{x \in \mathbb{R}^n_+ : x_n = 0\}$. For any $x, y \in \mathbb{R}^n_+$, let $R := x + \frac{x-\bar{x}}{|x-\bar{x}|}|x-y|$ and $S := R + (\bar{y} - \bar{x})$. Then, by the choice $\Gamma := \overline{xR} \cup \overline{RS} \cup \overline{Sy}$, we can easily verify that $\mathbb{R}^n_+$ is an (ε, ∞) domain.

3. For an infinite layer $\Omega := \{x \in \mathbb{R}^n : 0 < x_n < b\}$, where $b > 0$ denotes the layer's thickness, we proceed in a similar way. Let $x, y \in \Omega$, and $\bar{x}$ be the same projection as in the halfspace case. Then we define $R := \bar{x} + \frac{x-\bar{x}}{|x-\bar{x}|} \cdot \frac{b}{2}$. Taking $S := R + (\bar{y} - \bar{x})$ and $\Gamma := \overline{xR} \cup \overline{RS} \cup \overline{Sy}$, we see that infinite layers are (ε, ∞) domains, too.

4. Let $x' = (x_1, \ldots, x_{n-1})$ denote the first $n-1$ components of a vector in $\mathbb{R}^n$ and $\omega \in C^1_c(\mathbb{R}^{n-1})$. We define the *bent halfspace* by $H_\omega := \{x = (x', x_n) \in \mathbb{R}^n : x_n > \omega(x')\}$. Obviously, the mapping $\phi : H_\omega \to \mathbb{R}^n_+$, $\phi(x) := (x', x_n - \omega(x'))$ is bi-Lipschitz, i.e. ϕ and ϕ^{-1} are Lipschitz continuous. From [Jon81] we know the following result: If Ω is an (ε, ∞) domain and $\phi : \Omega \to \phi(\Omega)$ is bi-Lipschitz, then $\phi(\Omega)$ is an (ε, ∞) domain, too. Thus, H_ω is also an (ε, ∞) domain.

Chapter 2

The Stokes equations in $\mathbb{R}^n_+$

Some of the key results of this thesis rely on properties of the Stokes operator in the n-dimensional halfspace $\mathbb{R}^n_+ := \{x \in \mathbb{R}^n : x_n > 0\}$. For this reason, in this chapter we will survey the Stokes equations and the corresponding operator, acting in such domains.

In this section, we consider the Stokes equations in $\mathbb{R}^n_+$ given by

$$\left\{ \begin{array}{rcl} \partial_t u - \Delta u + \nabla p &=& 0, \quad t > 0,\ x \in \mathbb{R}^n_+, \\ \operatorname{div} u &=& 0, \quad t > 0,\ x \in \mathbb{R}^n_+, \\ u(t,x) &=& 0, \quad t > 0,\ x_n = 0,\ x' \in \mathbb{R}^{n-1}, \\ u(0,x) &=& u_0, \quad x \in \mathbb{R}^n_+, \end{array} \right. \tag{2.1}$$

as well as the corresponding Stokes resolvent equation

$$\left\{ \begin{array}{rcl} \lambda u - \Delta u + \nabla p &=& f, \quad x \in \mathbb{R}^n_+, \\ \operatorname{div} u &=& 0, \quad x \in \mathbb{R}^n_+, \\ u &=& 0, \quad x_n = 0,\ x' \in \mathbb{R}^{n-1}, \end{array} \right. \tag{2.2}$$

with a given right hand side $f \in L^p_\sigma(\mathbb{R}^n_+)$. In these equations $u : \mathbb{R}^n_+ \to \mathbb{R}^n$ denotes a vector field and $p : \mathbb{R}^n_+ \to \mathbb{R}$ is a scalar function.

Applying the Helmholtz decomposition to (2.1), we obtain the linear evolution equation

$$\left\{ \begin{array}{rcl} u'(t) + Au(t) &=& 0, \quad t > 0, \\ u(0) &=& u_0, \end{array} \right. \tag{2.3}$$

where A denotes the Stokes operator in $L^p_\sigma(\mathbb{R}^n_+)$, cf. Definition 1.36.

2.1 Resolvent estimates for the Stokes operator

The Stokes equations respectively the Stokes resolvent equations in halfspaces were first investigated in [Sol77], [McC81], and [Uka87]. In these articles the authors proved the following result.

Proposition 2.1. *Let $\theta \in (0,\pi)$ and $\lambda \in \Sigma_\theta$. Furthermore, let $f \in L^p_\sigma(\mathbb{R}^n_+)$. Then there exists a unique solution $(u,p) \in (W^{2,p}(\mathbb{R}^n_+)^n \cap W^{1,p}_0(\mathbb{R}^n_+)^n \cap L^p_\sigma(\mathbb{R}^n_+)) \times \widehat{W}^{1,p}(\mathbb{R}^n_+)$ of the Stokes resolvent equations (2.2). Moreover, for $\theta \in (0,\pi)$ and $\lambda \in \Sigma_\theta$ the resolvent estimate*

$$|\lambda|\|u\|_{L^p(\mathbb{R}^n_+)^n} + |\lambda|^{\frac{1}{2}}\|u\|_{\widehat{W}^{1,p}(\mathbb{R}^n_+)^n} + \|u\|_{\widehat{W}^{2,p}(\mathbb{R}^n_+)^n} + \|\nabla\pi\|_{L^p(\mathbb{R}^n_+)^n} \le C\|f\|_{L^p(\mathbb{R}^n_+)^n}$$

holds, where $C > 0$ is a constant depending on p, θ, and n only.

Let $(\lambda + A)^{-1}$ be the resolvent of the negative Stokes operator $-A$ in $L^p_\sigma(\mathbb{R}^n_+)$ at the point λ. Then, from the resolvent estimate given in Proposition 2.1 it follows that $-A$ generates a bounded analytic semigroup in $L^p_\sigma(\mathbb{R}^n_+)$.

Lateron, in [DHP01, Theorem 7.6] it was shown that A even admits an $\mathcal{R}$-bounded $\mathcal{H}^\infty$-calculus in $L^p_\sigma(\mathbb{R}^n_+)$ of angle 0. This result directly implies that A is also an $\mathcal{R}$-sectorial operator of angle 0 in $L^p_\sigma(\mathbb{R}^n_+)$. Thus, we have for all $\theta \in (0,\pi)$

$$\mathcal{R}_{L^p_\sigma(\mathbb{R}^n_+)}\{\lambda(\lambda + A)^{-1} : \lambda \in \Sigma_\theta\} \le C \tag{2.4}$$

with a constant $C > 0$, which may depend on θ. Having in mind the Weis characterization theorem of maximal L^p-regularity for general sectorial operators acting in UMD-Banach spaces (see Proposition 1.31), it follows from (2.4) that the Stokes operator A admits maximal regularity in $L^p_\sigma(\mathbb{R}^n_+)$. Hence, the results of [DHP01] imply the following proposition.

Proposition 2.2. *Let $1 < p,q < \infty$ and $f \in L^q((0,\infty), L^p_\sigma(\mathbb{R}^n_+))$. Then there exist a unique $u \in L^q((0,\infty), D(A))$ with $u' \in L^q((0,\infty), L^p_\sigma(\mathbb{R}^n_+))$ satisfying*

$$\begin{cases} u'(t) + Au(t) &= f(t), \quad t \in (0,\infty), \\ u(0) &= 0, \end{cases}$$

and a constant $C > 0$, such that

$$\|u'\|_{L^q((0,\infty),L^p_\sigma(\mathbb{R}^n_+))} + \|Au\|_{L^q((0,\infty),L^p_\sigma(\mathbb{R}^n_+))} \le C\|f\|_{L^q((0,\infty),L^p_\sigma(\mathbb{R}^n_+))}. \tag{2.5}$$

In the following we collect further properties of the operator $(\lambda + A)^{-1}$ concerning its $\mathcal{R}$-boundedness. By the contraction principle for $\mathcal{R}$-bounds (see for example [DHP03]), we obtain from inequality (2.4) that for every $\theta \in (0,\pi)$ there is a constant $C > 0$, such that

$$\mathcal{R}_{L^p_\sigma(\mathbb{R}^n_+)}\{(\lambda + A)^{-1} : \lambda \in \lambda_0 + \Sigma_\theta\} \le \frac{C}{\lambda_0}, \quad \lambda_0 > 0, \tag{2.6}$$

holds. Moreover, using the graph norm of A, for every $u \in D(A)$ we have

$$\|u\|_{W^{2,p}(\mathbb{R}^n_+)^n} \le C(\|Au\|_{L^p_\sigma(\mathbb{R}^n_+)} + \|u\|_{L^p_\sigma(\mathbb{R}^n_+)}). \tag{2.7}$$

Now, let $\theta \in (0,\pi)$ and $\lambda \in \Sigma_\theta$. Applying inequality (2.7) yields

$$\begin{aligned} &\|(\lambda+A)^{-1}\|_{\mathscr{L}(L^p_\sigma(\mathbb{R}^n_+),W^{2,p}(\mathbb{R}^n_+)^n)} \\ \leq\ & \|A(\lambda+A)^{-1}\|_{\mathscr{L}(L^p_\sigma(\mathbb{R}^n_+))} + \|(\lambda+A)^{-1}\|_{\mathscr{L}(L^p_\sigma(\mathbb{R}^n_+))} \\ \leq\ & \|\mathrm{Id}-\lambda(\lambda+A)^{-1}\|_{\mathscr{L}(L^p_\sigma(\mathbb{R}^n_+))} + \|(\lambda+A)^{-1}\|_{\mathscr{L}(L^p_\sigma(\mathbb{R}^n_+))} \\ \leq\ & C\left(2+\frac{1}{|\lambda|}\right). \end{aligned}$$

By the calculation rules for $\mathcal{R}$-bounds, this implies that for given $\theta \in (0,\pi)$ and $\lambda_0 > 0$ there is a $C > 0$, such that

$$\mathcal{R}_{L^p_\sigma(\mathbb{R}^n_+)\to W^{2,p}(\mathbb{R}^n_+)^n}\{(\lambda+A)^{-1} : \lambda \in \lambda_0+\Sigma_\theta\} \leq C.$$

In particular, we have for $\theta \in (0,\pi)$

$$\mathcal{R}_{L^p_\sigma(\mathbb{R}^n_+)\to W^{2,p}(\mathbb{R}^n_+)^n}\{(\lambda+A)^{-1} : \lambda \in 1+\Sigma_\theta\} \leq C. \tag{2.8}$$

Combining Lemma 1.27 with the above inequalities (2.6) and (2.8), we immediately obtain the following result on the $\mathcal{R}$-boundedness of $\{(\lambda+A)^{-1} : \lambda \in \lambda_0+\Sigma_\theta\}$.

Lemma 2.3. *Let $1 < p < \infty$ and $\theta \in (0,\pi)$. Then, for every $s \in [0,2]$ there is a $C > 0$, such that*

$$\lambda_0^{\frac{2-s}{2}}\mathcal{R}_{L^p_\sigma(\mathbb{R}^n_+)\to W^{s,p}(\mathbb{R}^n_+)^n}\{(\lambda+A)^{-1} : \lambda \in \lambda_0+\Sigma_\theta\} \leq C, \quad \lambda_0 \geq 1,$$

holds.

Remark 2.4. We point out that Lemma 2.3 may be transferred to more general domains than the halfspace. Actually, besides Lemma 1.27, it is based on the assertions (2.4) and (2.7) only. Hence, the assertion of Lemma 2.3 remains true for a domain Ω instead of $\mathbb{R}^n_+$, whenever (2.4) and (2.7) hold true for Ω.

2.2 A decay estimate for the pressure term

In this section we will prove a result concerning the decay of the pressure term p in (2.2) in terms of $\mathcal{R}$-bounds. As a preparation, we state an interpolation property of the Dirichlet Laplacian, which is taken from [NS03].

Proposition 2.5. *Let $1 < p < \infty$ and let Ω be a domain of class C^3, which is either bounded, exterior, $\mathbb{R}^n$, or a perturbed halfspace. Let Δ denote the Dirichlet Laplacian in $L^p(\Omega)$ with domain $D(\Delta) = W^{2,p}(\Omega)\cap W^{1,p}_0(\Omega)$. Then we have for $\alpha \in (0,\frac{1}{2p})$*

$$[L^p(\Omega), D(\Delta)]_\alpha = W^{2\alpha,p}(\Omega).$$

For a proof, we refer to [NS03].

Having the foregoing result at hands, we are able to prove the following lemma concerning the associated pressure, which generalizes the result given in [NS03, Lemma 13] to the context of $\mathcal{R}$-bounds.

Lemma 2.6. *For $n \geq 2$ consider a bounded domain $G \subset \mathbb{R}^n_+$ with Lipschitz boundary. Furthermore, let $1 < p < \infty$, $\theta \in (0, \pi)$, and $\lambda_0 \geq 1$. Then for $\lambda \in \lambda_0 + \Sigma_\theta$ we denote by $(u_\lambda, \pi_\lambda) \in D(A) \times \widehat{W}^{1,p}(\mathbb{R}^n_+)$ the unique solution of (2.2) satisfying $\int_G \pi_\lambda \, dx = 0$. Moreover, we define the operator $\Pi_\lambda : L^p_\sigma(\mathbb{R}^n_+) \to L^p(G)$ by*

$$\Pi_\lambda f := \pi_\lambda|_G.$$

Then for every $\alpha \in (0, \frac{1}{2p'})$ and all $s \in [0,1]$ there exists $C > 0$, such that

$$\lambda_0^{(1-s)\alpha} \mathcal{R}_{L^p_\sigma(\mathbb{R}^n_+) \to W^{s,p}(G)} \{\Pi_\lambda : \lambda \in \lambda_0 + \Sigma_\theta\} \leq C, \quad \lambda_0 \geq 1. \tag{2.9}$$

Proof. Let $G \subset \mathbb{R}^n_+$ be a bounded domain with Lipschitz boundary, $N \in \mathbb{N}$, $\alpha \in (0, \frac{1}{2p'})$, $\lambda_j \in \Sigma_\theta$, $f_j \in L^p_\sigma(\mathbb{R}^n_+)$, and ε_j independent, symmetric $\{-1,1\}$-valued random variables on a probability space $(\mathcal{M}, \mathcal{A}, \mu)$, $j = 1, \ldots, N$. For $\varphi \in L^{p'}_0(G)$ let $\phi := B_G \varphi \in W^{1,p'}_0(G)^n$ and $\tilde{\phi}$ its extension to $\mathbb{R}^n_+$ by 0. We define $\pi_j := \Pi_{\lambda_j} f_j$. By integration by parts and the properties of the Helmholtz projection (see Proposition 1.34) we may calculate

$$\begin{aligned} \langle \pi_j, \varphi \rangle_G &= \langle \pi_j, \operatorname{div} \phi \rangle_G = -\langle \nabla \pi_j, \phi \rangle_G = -\langle \nabla \pi_j, \tilde{\phi} \rangle_{\mathbb{R}^n_+} \\ &= -\langle (\mathrm{Id} - P) \Delta u_{\lambda_j}, \tilde{\phi} \rangle_{\mathbb{R}^n_+} = \langle -\Delta u_{\lambda_j}, (\mathrm{Id} - P) \tilde{\phi} \rangle_{\mathbb{R}^n_+}. \end{aligned}$$

Since $-\Delta$ admits an $\mathcal{R}$-bounded $\mathcal{H}^\infty$-calculus in $L^p(\mathbb{R}^n_+)^n$ (see for example [DHP01], [DHP03], or [KW04]) and so has bounded imaginary powers on $L^p(\mathbb{R}^n_+)^n$, according to Propositions 1.24 and 2.5 we have $D((-\Delta)^\alpha) = W^{2\alpha,p'}(\mathbb{R}^n_+)^n$. Furthermore, $L^p_0(G)' = L^{p'}_0(G)$ holds (see [GHH06b]), and we have $P \in \mathscr{L}(W^{1,r}(\mathbb{R}^n_+))$ for $1 < r < \infty$ (see [NS03, Proposition 20]). Moreover, by assumption we have $\alpha < \frac{1}{2}$. Hence, the estimate

$$\|(-\Delta)^\alpha (\mathrm{Id} - P) \tilde{\phi}\|_{L^{p'}(\mathbb{R}^n_+)^n} \leq C \|B_G \varphi\|_{W^{1,p'}(G)^n} \leq C \|\varphi\|_{L^{p'}(G)}$$

follows. This yields

$$\begin{aligned}
&\Big\|\sum_{j=1}^{N}\varepsilon_j(\omega)\pi_j\Big\|_{L^p(G)}\\
=&\sup_{\varphi\in L_0^{p'}(G),\|\varphi\|_{L^{p'}(G)}=1}\Big|\Big\langle\sum_{j=1}^{N}\varepsilon_j(\omega)\pi_j,\varphi\Big\rangle_G\Big|\\
=&\sup_{\varphi\in L_0^{p'}(G),\|\varphi\|_{L^{p'}(G)}=1}\Big|\sum_{j=1}^{N}\varepsilon_j(\omega)\langle-\Delta u_j,(\mathrm{Id}-P)\tilde{\phi}\rangle\Big|\\
=&\sup_{\varphi\in L_0^{p'}(G),\|\varphi\|_{L^{p'}(G)}=1}\Big|\sum_{j=1}^{N}\varepsilon_j(\omega)\langle(-\Delta)^{1-\alpha}u_j,(-\Delta)^{\alpha}(\mathrm{Id}-P)\tilde{\phi}\rangle\Big|\\
\leq&\sup_{\varphi\in L_0^{p'}(G),\|\varphi\|_{L^{p'}(G)}=1}C\Big\|\sum_{j=1}^{N}\varepsilon_j(\omega)(-\Delta)^{1-\alpha}u_j\Big\|_{L^p(\mathbb{R}^n_+)^n}\|\tilde{\phi}\|_{W^{1,p'}(\mathbb{R}^n_+)^n}\\
\leq&\ C\Big\|\sum_{j=1}^{N}\varepsilon_j(\omega)(-\Delta)^{1-\alpha}u_j\Big\|_{L^p(\mathbb{R}^n_+)^n}.
\end{aligned}$$

Considering the function $f_\lambda(z)=\frac{z^{1-\alpha}}{\lambda-z}$, $z\in\Sigma_{\pi-\theta-\varepsilon}$ for suitable $\varepsilon>0$ and applying the $\mathcal{R}$-bounded $\mathcal{H}^\infty$-calculus of $-\Delta$ to the family $\{f_\lambda:-\lambda\in\lambda_0+\Sigma_\theta\}$ we obtain

$$\lambda_0^{\alpha}\mathcal{R}_{L^p(\mathbb{R}^n_+)^n}\{(-\Delta)^{1-\alpha}R(\lambda,\Delta):\lambda\in\lambda_0+\Sigma_\theta\}\leq C,\quad\lambda_0>0.\tag{2.10}$$

Then, putting $s=0$ in Lemma 2.3 and using the representation $u_j=R(\lambda_j,\Delta)f_j-R(\lambda_j,\Delta)(\mathrm{Id}-P)\Delta(\lambda_j+A)^{-1}f_j$, we obtain from inequality (2.10) that

$$\int_{\mathcal{M}}\Big\|\sum_{j=1}^{N}\varepsilon_j(\omega)(-\Delta)^{1-\alpha}u_j\Big\|_{L^p(\mathbb{R}^n_+)^n}\,\mathrm{d}\omega\leq C\lambda_0^{-\alpha}\int_{\mathcal{M}}\Big\|\sum_{j=1}^{N}\varepsilon_j(\omega)f_j\Big\|_{L^p(\mathbb{R}^n_+)^n}\,\mathrm{d}\omega,$$

which implies the assertion for $s=0$. Moreover, putting $s=2$ in Lemma 2.3 and using the representation $\nabla\pi_j=(\mathrm{Id}-P)\Delta(\lambda_j+A)^{-1}f_j$, we have

$$\begin{aligned}
&\int_{\mathcal{M}}\Big\|\sum_{j=1}^{N}\varepsilon_j(\omega)\nabla\pi_j\Big\|_{L^p(G)}\,\mathrm{d}\omega\\
\leq&\int_{\mathcal{M}}\Big\|\sum_{j=1}^{N}\varepsilon_j(\omega)(\mathrm{Id}-P)\Delta(\lambda_j+A)^{-1}f_j\Big\|_{L^p(\mathbb{R}^n_+)^n}\,\mathrm{d}\omega\\
\leq&\ C\int_{\mathcal{M}}\Big\|\sum_{j=1}^{N}\varepsilon_j(\omega)f_j\Big\|_{L^p(\mathbb{R}^n_+)^n}\,\mathrm{d}\omega,
\end{aligned}$$

which together with the foregoing arguments gives the assertion for $s = 1$. Now, the interpolation result of Lemma 1.27 yields the claim for all $s \in (0, 1)$. □

Remark 2.7. The assertion of Lemma 2.6 with $\mathbb{R}^n_+$ replaced by Ω holds true provided (2.4) and (2.7) hold true for Ω.

Chapter 3

Maximal regularity of the Stokes equations

This chapter deals with the Stokes equations

$$\left\{\begin{array}{rcll} \partial_t u - \Delta u + \nabla p & = & 0, & t > 0,\ x \in \Omega, \\ \operatorname{div} u & = & 0, & t > 0,\ x \in \Omega, \\ u(t,x) & = & 0, & t > 0,\ x \in \partial\Omega, \\ u(0,x) & = & u_0, & x \in \Omega, \end{array}\right. \tag{3.1}$$

for C^3-domains $\Omega \subset \mathbb{R}^n$ with compact boundary, i.e. Ω is either bounded or an exterior domain. Regarding (3.1) as the linearization of the nonlinear Navier-Stokes equations, a deep knowledge of that system may be helpful for the investigation of the latter ones. In particular, the property of maximal L^p-regularity of the linear system provides an approach to construct solutions to the nonlinear one using fixed point arguments (see for example [Ama95]). In order to apply the concept of maximal regularity as introduced in Section 1.2 we need to transform system (3.1) into an abstract Cauchy problem. For this purpose, we apply the Helmholtz projection P to the Stokes equations and obtain the evolution equation

$$\left\{\begin{array}{rcll} u'(t) + Au(t) & = & 0, & t \in J, \\ u(0) & = & u_0, & \end{array}\right. \tag{3.2}$$

in $L^p_\sigma(\Omega)$ for some time interval $J = (0,T)$ with $0 < T \leq \infty$. Then we say that the Stokes equations (3.1) have maximal regularity, if the according operator A does.

In view of Proposition 1.31 the property of the Stokes equations to have maximal regularity is equivalent to A being $\mathcal{R}$-sectorial in $L^p_\sigma(\Omega)$ with $\mathcal{R}$-angle $\phi_A^{\mathcal{R}} < \frac{\pi}{2}$. Hence, we proceed considering the Stokes resolvent equations

$$\lambda u + Au = f, \quad x \in \Omega, \tag{3.3}$$

in Ω with some function $f \in L^p_\sigma(\Omega)$. Starting with system (3.3) we will end up with the result that $\{\lambda(\lambda + A)^{-1} : \lambda \subset \lambda_0 + \Sigma_\theta\}$ is an $\mathcal{R}$-bounded subset of $\mathscr{L}(L^p_\sigma(\Omega))$ for some $\lambda_0 \geq 0$ and any $\theta \in (0, \pi)$. That means that the (possibly shifted) Stokes operator is $\mathcal{R}$-sectorial of angle 0.

The essential step in the proof of our result will be a localization procedure, which transfers the well-known results for the halfspace case (see Chapter 2) to the case of a bounded domain, whose boundary is of class C^3. For this purpose, we choose a partition of unity $(\varphi_j)_{j\in\{1,\dots,N\}}$ subordinated to a finite covering of Ω and transform (3.3) on $\Omega \cap \operatorname{supp} \varphi_j$ to the corresponding problem in $\mathbb{R}^n_+$.

The major difficulty in using localization methods for the Stokes equation is to preserve the solenoidality of the obtained solutions. Actually, the usual procedure for elliptic problems (see for example [ADN59]) fails to transform divergence free vector fields on the halfspace into the space of divergence free vector fields on a bounded domain. To handle this problem the Bogovskiĭ operator comes into play. More precisely, we construct a function u as

$$u := \sum_{j=1}^{N} (\varphi_j u_j - B((\nabla \varphi_j) u_j)), \tag{3.4}$$

where φ_j are the cut-off functions coming from the partition of unity, u_j are solutions to the restricted problem on suitable sets Ω_j, and B denote Bogovskiĭ's operator on Ω.

In [SS03] Shibata and Shimizu used a localization procedure to prove estimates for the solution of a generalized Stokes resolvent equation with Neumann boundary conditions in bounded or exterior domains. Since they deal with the generalized Stokes system, they do not have to assure that the solution remains divergence free under the according transformations.

When applying localization methods, there will appear several correction terms, which entail that the constructed function u is not a solution to the original problem (3.3). The exact solution is therefore obtained by a Neumann series argument due to several operator estimates. The exterior domain case will then be deduced from the according results for bounded domains and the whole space $\mathbb{R}^n$.

The property of the Stokes operator A to admit maximal regularity for $1 < p < \infty$ was first proved by Solonnikov in [Sol77] by the use of potential theory. Further proofs of this fact were given by Giga and Sohr [GS91a], by Grubb and Solonnikov [GS91b], using pseudo-differential techniques, and by Fröhlich [Frö07], making use of the concept of weighted estimates with respect to Muckenhoupt weights.

3.1 Main results

In this section we state the main results of this chapter. As mentioned before, to show maximal regularity of the Stokes operator we use the characterization in terms of $\mathcal{R}$-boundedness as given in Proposition 1.31. In particular, we prove the following result:

Theorem 3.1. *Let $\Omega \subset \mathbb{R}^n$ be a domain with compact boundary $\partial\Omega$ of class C^3 and $1 < p < \infty$. Then there is an $\omega \geq 0$, such that $A+\omega$ is an $\mathcal{R}$-sectorial operator in $L^p_\sigma(\Omega)$ of angle $\phi^{\mathcal{R}}_A = 0$. If Ω is bounded, we have $0 \in \rho(A)$. In this case we may choose $\omega = 0$.*

In view of Proposition 1.31, Theorem 3.1 implies that the Stokes operator A has maximal regularity for $1 < p < \infty$. Hence, as a corollary we obtain the following result, which was proved first by Solonnikov (see [Sol77]).

Corollary 3.2. *Let $1 < p, q < \infty$, $J = (0, T)$ for some $T > 0$ and $\Omega \subset \mathbb{R}^n$ be a domain as above. Then for every $f \in L^q(J, L^p_\sigma(\Omega))$ there exist a unique $u \in L^q(J, D(A)) \cap W^{1,q}(J, L^p_\sigma(\Omega))$ satisfying*

$$\begin{cases} u'(t) + Au(t) &= f(t), \quad t \in J, \\ u(0) &= 0, \end{cases}$$

and some $C > 0$, such that

$$\|u'\|_{L^q(J,L^p_\sigma(\Omega))} + \|Au\|_{L^q(J,L^p_\sigma(\Omega))} \leq C\|f\|_{L^q(J,L^p_\sigma(\Omega))}.$$

If Ω is bounded, then we may choose $J = (0, \infty)$.

By Theorem 3.1, there is $\lambda_0 \geq 0$, such that for $f \in L^p_\sigma(\Omega)$ the resolvent equation

$$\lambda u + Au = f$$

has a unique solution $u \in D(A)$ for all $\lambda \in \lambda_0 + \Sigma_\theta$. The solution u is given by $u = (\lambda + A)^{-1} f$. Now, setting $\nabla p := (\mathrm{Id} - P)\Delta u$ we also obtain a solution of the Stokes resolvent problem (3.3). In particular, we have the following result.

Corollary 3.3. *Let $1 < p < \infty$, $\Omega \subset \mathbb{R}^n$ as above and let $\theta \in (0, \pi)$. Then there exists $\lambda_0 \in \mathbb{R}$, such that for all $\lambda \in \lambda_0 + \Sigma_\theta$ and $f \in L^p_\sigma(\Omega)$ there exists a unique $(u, p) \in (W^{2,p}(\Omega)^n \cap W^{1,p}_0(\Omega)^n \cap L^p_\sigma(\Omega)) \times \widehat{W}^{1,p}(\Omega)$ satisfying (3.3). Moreover, there exists $C > 0$, such that for all $\lambda \in \lambda_0 + \Sigma_\theta$ and $f \in L^p_\sigma(\Omega)$*

$$|\lambda - \lambda_0| \|u\|_{L^p(\Omega)^n} + \|\nabla^2 u\|_{L^p(\Omega)^n} + \|\nabla p\|_{L^p(\Omega)^n} \leq C\|f\|_{L^p(\Omega)^n}.$$

If Ω is bounded, then λ_0 can be chosen strictly negative.

Remark 3.4. By Remark 1.18 we immediately obtain that for $1 < p < \infty$ the operator $-(A + \lambda_0)$ is the generator of a bounded analytic C_0-semigroup on $L^p_\sigma(\Omega)$. By Remark 1.15 this implies that the negative Stokes operator $-A$ generates an analytic semigroup T on $L^p_\sigma(\Omega)$. If Ω is bounded, then the growth bound of T is negative, which means that there exist constants $M, \delta > 0$, such that $\|T(t)\| \leq Me^{-\delta t}$ for $t \geq 0$.

3.2 Change of coordinates

In this chapter we deal with the Stokes resolvent problem on domains of class C^3, which are either bounded or exterior. The strategy of handling such domains will be a localization method, which allows us to reduce the problem to a similar problem in the halfspace. For this purpose we introduce so-called local coordinates. For a point $x_0 \in \partial\Omega$ consider the shift $s : \Omega \to \Omega'$ with $s(x) = x - x_0$. Furthermore, consider the rotation $r : \Omega' \to \tilde{\Omega}$ mapping the interior normal at x_0 into the direction of the positive x_n-axis. By the described procedure we obtain *local coordinates corresponding to* x_0.

Note that the map r is an orthogonal transformation in $\mathbb{R}^n$. Hence, transforming a Stokes resolvent problem on a given domain $\Omega \subset \mathbb{R}^n$ into an according problem in local coordinates corresponding to some $x_0 \in \partial\Omega$ leads to an equivalent Stokes resolvent problem. For example, for an open subset $Q \subset \mathbb{R}^n$ let (U, P) be a solution of the Stokes resolvent problem

$$\begin{cases} \lambda U - \Delta U + \nabla P &= F \quad \text{in } Q, \\ \operatorname{div} U &= 0 \quad \text{in } Q, \\ U &= 0 \quad \text{on } \partial Q, \end{cases}$$

for a given right hand side F. If $\mathcal{O}$ is an orthogonal transformation in $\mathbb{R}^n$ and $\tilde{x} := V\tilde{x} := \mathcal{O}(x - x_0)$, then $\tilde{U}(\tilde{x}) := \mathcal{O}U(V^{-1}\tilde{x})$ and $\tilde{P}(\tilde{x}) := P(V^{-1}\tilde{x})$ solve the equivalent Stokes resolvent problem on VQ with right hand side $\tilde{F}(\tilde{x}) := \mathcal{O}F(V^{-1}\tilde{x})$. For this reason, to simplify the notation these transformations will be omitted in this chapter.

Throughout this chapter $\Omega \subset \mathbb{R}^n$ denotes a bounded C^3-domain. In the following we assume $x_0 \in \partial\Omega$ and choose local coordinates corresponding to x_0. By definition of a C^3-boundary, there exists an open neighbourhood $U = U_1 \times U_2 \subset \mathbb{R}^n$ containing $x_0 = 0$ with $U_1 \subset \mathbb{R}^{n-1}$ and $U_2 \subset \mathbb{R}$ open and a function $h \in C^3(\overline{U_1})$ satisfying $\partial\Omega \cap U = \{x = (x', x_n) \in U : x_n = h(x')\}$ and $\Omega \cap U = \{x \in U : x_n > h(x')\}$. Setting

$$g(x) := \begin{pmatrix} x' \\ x_n - h(x') \end{pmatrix}, \quad x \in U, \tag{3.5}$$

we obtain an injection $g \in C^3(\overline{U}, \mathbb{R}^n)$ satisfying $\Omega \cap U = \{x \in U : g_n(x) > 0\}$ and $\partial\Omega \cap U = \{x \in U : g_n(x) = 0\}$. By compactness of $\partial\Omega$, all derivatives of g and of g^{-1} (defined on $\hat{U} := g(U)$) up to order 3 may be assumed to be bounded by a constant independent of x_0. In particular, take a small enough, such that $Q_{x_0,a} \subset U$, where $Q_{x_0,a}$ is an open cube with center $x_0 \in \partial\Omega$ of sidelength a. If we set $\hat{Q}_{x_0,a} := g_{x_0,a}(Q_{x_0,a})$ with

$$g_{x_0,a}(x) := \begin{pmatrix} x' \\ x_n - h_{x_0,a}(x') \end{pmatrix}, \quad x \in Q_{x_0,a},$$

for some appropriate function $h_{x_0,a}$, then by the regularity assumption on $\partial\Omega$, there exist constants $C, b_1 > 0$, such that

$$\|h_{x_0,a}\|_{W^{3,\infty}(\mathbb{R}^{n-1})} \leq C, \quad a \in (0, b_1), \; x_0 \in \partial\Omega, \tag{3.6}$$

and for $\varepsilon > 0$ there exists $a_\varepsilon > 0$, such that

$$\|\nabla h_{x_0,a}\|_{L^\infty(\mathbb{R}^{n-1})^{n-1}} \leq \varepsilon, \quad a \in (0, a_\varepsilon), \; x_0 \in \partial\Omega. \tag{3.7}$$

In the following we will turn to several transformation mappings, which allow to transfer subdomains of a C^3-domain Ω to suitable subsets of the halfspace $\mathbb{R}^n_+$.

Definition 3.5. Let $x_0 \in \partial\Omega$ and let $a > 0$, such that (3.6) is fulfilled.

1. Consider a function $u\colon Q_{x_0,a} \cap \Omega \to \mathbb{R}$. Then we define the push-forward $v = \mathcal{G}_{x_0,a}u$ on $\hat{Q}_{x_0,a} \cap \mathbb{R}^n_+$ by $v(y) := u(g^{-1}_{x_0,a}(y))$.

   Analogously, for a function $v\colon \hat{Q}_{x_0,a} \cap \mathbb{R}^n_+ \to \mathbb{R}$ the pull-back $u = \mathcal{G}^{-1}_{x_0,a}v$ defined on $Q_{x_0,a} \cap \Omega$ is given by $u(x) := v(g_{x_0,a}(x))$.

2. Consider a vector field $u\colon Q_{x_0,a} \cap \Omega \to \mathbb{R}^n$. Then the according push-forward $v_\sigma = \mathcal{G}_{\sigma,x_0,a}u$ defined on $\hat{Q}_{x_0,a} \cap \mathbb{R}^n_+$ is given by $v_\sigma(y) := J_{g_{x_0,a}}(g^{-1}_{x_0,a}(y)) \cdot (u(g^{-1}_{x_0,a}(y)))$, where $J_{g_{x_0,a}}$ denotes the Jacobian of $g_{x_0,a}$.

   Note that $J_{g^{-1}_{x_0,a}} = J^{-1}_{g_{x_0,a}}$. Hence, for a vector field $v\colon \hat{Q}_{x_0,a} \cap \mathbb{R}^n_+ \to \mathbb{R}^n$, we define the corresponding pull-back $u_\sigma = \mathcal{G}^{-1}_{\sigma,x_0,a}v$ on $Q_{x_0,a} \cap \Omega$ by $u_\sigma(x) := J^{-1}_{g_{x_0,a}}(g_{x_0,a}(x)) \cdot (v(g_{x_0,a}(x)))$.

Lemma 3.6. *Consider the linear transformations $\mathcal{G}_{x_0,a}$ and $\mathcal{G}_{\sigma,x_0,a}$ as given in Definition 3.5.*

1. The transformation $\mathcal{G}_{x_0,a}$ induces isomorphisms $\mathcal{G}^{(s,p)}_{x_0,a}\colon W^{s,p}(Q_{x_0,a} \cap \Omega) \to W^{s,p}(\hat{Q}_{x_0,a} \cap \mathbb{R}^n_+)$ for $p \in (1,\infty)$ and $s \in [-2,2]$.

2. The transformation $\mathcal{G}_{\sigma,x_0,a}$ induces isomorphisms $\mathcal{G}^{(s,p)}_{\sigma,x_0,a}\colon L^p_\sigma(Q_{x_0,a} \cap \Omega) \to L^p_\sigma(\hat{Q}_{x_0,a} \cap \mathbb{R}^n_+)$ and $\mathcal{G}^{(s,p)}_{\sigma,x_0,a}\colon W^{s,p}(Q_{x_0,a} \cap \Omega)^n \to W^{s,p}(\hat{Q}_{x_0,a} \cap \mathbb{R}^n_+)^n$ for $p \in (1,\infty)$ and $s \in [-2,2]$.

Proof. 1. We start by proving that $\mathcal{G}^{(s,p)}_{x0,a} \in \mathscr{L}(W^{s,p}(Q_{x0,a} \cap \Omega), W^{s,p}(\hat{Q}_{x0,a} \cap \mathbb{R}^n_+)$ for every $s \in [-2,2]$ and all $1 < p < \infty$.

First we consider the case $s = 0$. Then for $f \in L^p(Q_{x0,a} \cap \Omega)$ by the transformation formula

$$\begin{aligned}
\|\mathcal{G}^{(s,p)}_{x0,a} f\|_{L^p(\hat{Q}_{x_0,a} \cap \mathbb{R}^n_+)} &\leq \|J_{g_{x_0,a}}\| \|f \circ g^{-1}_{x0,a}\|_{L^p(\hat{Q}_{x_0,a} \cap \mathbb{R}^n_+)} \\
&= \|J_{g_{x_0,a}}\| \| |f| \det J_{g_{x_0,a}} \|_{L^p(Q_{x_0,a} \cap \Omega)} \\
&\leq C(1 + \|h_{x_0,a}\|_{W^{3,\infty}(\mathbb{R}^{n-1})}) \|f\|_{L^p(Q_{x_0,a} \cap \Omega)}
\end{aligned}$$

holds.

Next we turn to the case $s = 2$. Combining the chain rule with the transformation formula we obtain

$$\begin{aligned}
&\|\nabla \mathcal{G}^{(s,p)}_{x0,a} f\|_{L^p(\hat{Q}_{x_0,a} \cap \mathbb{R}^n_+)} \\
= \;& \|\nabla (f \circ g^{-1}_{x0,a})\|_{L^p(\hat{Q}_{x_0,a} \cap \mathbb{R}^n_+)} \\
\leq \;& C(n)(1 + \|h_{x_0,a}\|_{W^{3,\infty}(\mathbb{R}^{n-1})}) \|(\nabla f) \circ g^{-1}_{x0,a})\|_{L^p(\hat{Q}_{x_0,a} \cap \mathbb{R}^n_+)} \\
= \;& C(n)(1 + \|h_{x_0,a}\|_{W^{3,\infty}(\mathbb{R}^{n-1})}) \| (\nabla f) | \det J_{g_{x_0,a}} \|_{L^p(Q_{x_0,a} \cap \Omega)} \\
\leq \;& C(n)(1 + \|h_{x_0,a}\|_{W^{3,\infty}(\mathbb{R}^{n-1})})^2 \|f\|_{W^{1,p}(Q_{x_0,a} \cap \Omega)}
\end{aligned}$$

and, using the same calculations,

$$\begin{aligned}
&\|\nabla^2 \mathcal{G}^{(s,p)}_{x0,a} f\|_{L^p(\hat{Q}_{x_0,a} \cap \mathbb{R}^n_+)} \\
\leq \;& C(n)(1 + \|h_{x_0,a}\|_{W^{3,\infty}(\mathbb{R}^{n-1})})^2 \| \nabla^2 f | \det J_{g_{x_0,a}} \|_{L^p(Q_{x_0,a} \cap \Omega)} \\
& + C(n)(1 + \|h_{x_0,a}\|_{W^{3,\infty}(\mathbb{R}^{n-1})}) \| \nabla f | \det J_{g_{x_0,a}} \|_{L^p(Q_{x_0,a} \cap \Omega)} \\
\leq \;& C(n)(1 + \|h_{x_0,a}\|_{W^{3,\infty}(\mathbb{R}^{n-1})})^3 \|f\|_{W^{2,p}(Q_{x_0,a} \cap \Omega)}.
\end{aligned}$$

This implies the estimate

$$\|\mathcal{G}^{(s,p)}_{x0,a} f\|_{W^{2,p}(\hat{Q}_{x_0,a} \cap \mathbb{R}^n_+)} \leq C \|f\|_{W^{2,p}(Q_{x_0,a} \cap \Omega)},$$

where the constant C may depend on $h_{x_0,a}$.

These estimates yield that $\mathcal{G}^{(s,p)}_{x0,a} \in \mathscr{L}(W^{s,p}(Q_{x_0,a} \cap \Omega), W^{s,p}(\hat{Q}_{x_0,a} \cap \mathbb{R}^n_+)$ for all $1 < p < \infty$ and for $s \in \{0,2\}$. The corresponding result for $s \in (0,2)$ follows by interpolation.

Now, let $s \in [-2,0)$ and $f \in W^{s,p}(Q_{x_0,a} \cap \Omega)$. According to Definition 1.3, f

is contained in $\left(W_0^{-s,p'}(Q_{x0,a}\cap\Omega)\right)'$. This yields

$$\begin{aligned}
&\|\mathcal{G}^{(s,p)}_{x0,a}f\|_{W^{s,p}(\hat{Q}_{x0,a}\cap\mathbb{R}^n_+)}\\
\leq\ & \|J_{g_{x0,a}}\|\|f\circ g^{-1}_{x0,a}\|_{W^{s,p}(\hat{Q}_{x0,a}\cap\mathbb{R}^n_+)}\\
\leq\ & C \sup_{v\in C_c^\infty(\hat{Q}_{x0,a}\cap\mathbb{R}^n_+),\|v\|_{W^{-s,p'}(\hat{Q}_{x0,a}\cap\mathbb{R}^n_+)}=1} |\langle f\circ g^{-1}_{x0,a}, v\rangle|\\
=\ & C \sup_{v\in C_c^\infty(\hat{Q}_{x0,a}\cap\mathbb{R}^n_+),\|v\|_{W^{-s,p'}(\hat{Q}_{x0,a}\cap\mathbb{R}^n_+)}=1} |\langle f, v\circ g_{x0,a}\rangle|\\
\leq\ & C\|f\|_{W^{s,p}(Q_{x0,a}\cap\Omega)}\|v\circ g_{x0,a}\|_{W^{-s,p'}(Q_{x0,a}\cap\Omega)}\\
\leq\ & C(\|g\|_{W^{-s,\infty}(Q_{x0,a}\cap\Omega)}+1)\|f\|_{W^{s,p}(Q_{x0,a}\cap\Omega)}.
\end{aligned}$$

Hence, we have shown that $\mathcal{G}^{(s,p)}_{x0,a}\in\mathscr{L}(W^{s,p}(Q_{x0,a}\cap\Omega),W^{s,p}(\hat{Q}_{x0,a}\cap\mathbb{R}^n_+))$ for every $s\in[-2,2]$ and $1<p<\infty$. Since the pull-back ${\mathcal{G}^{(s,p)}_{x0,a}}^{-1}$ yields directly the corresponding inverse mapping, we obtain that $\mathcal{G}^{(s,p)}_{x0,a}$ is bijective, which completes the proof.

2. Using the same techniques as above, we may show that for every $s\in[-2,2]$ and all $1<p<\infty$ the operator $\mathcal{G}_{\sigma,x0,a}$ induces an isomorphism $\mathcal{G}^{(s,p)}_{\sigma,x0,a}\colon W^{s,p}(Q_{x0,a}\cap\Omega)^n\to W^{s,p}(\hat{Q}_{x0,a}\cap\mathbb{R}^n_+)^n$.

It remains to show that $\mathcal{G}^{(s,p)}_{\sigma,x0,a}$ in fact maps $L^p_\sigma(Q_{x0,a}\cap\Omega)$ to $L^p_\sigma(\hat{Q}_{x0,a}\cap\mathbb{R}^n_+)$. Therefore, we assume $u\in L^p_\sigma(Q_{x0,a}\cap\Omega)$. Then the calculation

$$\begin{aligned}
\operatorname{div}\mathcal{G}^{(s,p)}_{\sigma,x0,a}u &= \operatorname{div}J_{g_{x0,a}}(g^{-1}_{x0,a}(y))\cdot(u(g^{-1}_{x0,a}(y)))\\
&= \sum_{i=1}^n(\nabla u_i)(g^{-1}_{x0,a}(y))\cdot\partial_i(g^{-1}_{x0,a}(y))\\
&\quad-\sum_{i=1}^{n-1}(\partial_i h_{x0,a})(g^{-1}_{x0,a}(y))\cdot\partial_n(u_i(g^{-1}_{x0,a}(y)))\\
&= \sum_{i=1}^n(\partial_i u_i)(g^{-1}_{x0,a}(y))+\sum_{i=1}^{n-1}(\partial_n u_i)(g^{-1}_{x0,a}(y))\cdot(\partial_i h_{x0,a})(g^{-1}_{x0,a}(y))\\
&\quad-\sum_{i=1}^{n-1}(\partial_i h_{x0,a})(g^{-1}_{x0,a}(y))\cdot(\partial_n u_i)(g^{-1}_{x0,a}(y))\\
&= (\operatorname{div}u)(g^{-1}_{x0,a}(y))\\
&= 0
\end{aligned}$$

shows that $\mathcal{G}^{(s,p)}_{\sigma,x0,a}$ keeps the solenoidality of u. Since $\mathcal{G}^{(s,p)}_{\sigma,x0,a}$ preserves the no slip condition on the boundary of $Q_{x0,a}\cap\Omega$, we also have

$$\nu\cdot\mathcal{G}^{(s,p)}_{\sigma,x0,a}u|_{\partial\hat{Q}_{x0,a}\cap\mathbb{R}^n_+}=0,$$

where ν denotes the outer normal unit vector on $\partial\hat{Q}_{x_0,a} \cap \mathbb{R}^n_+$. Taking into account [Gal94a, Lemma III 2.1] this proves the assertion. □

Remark 3.7. In the sequel we will omit the superscript (s,p) and simply write $\mathcal{G}_{\sigma,x_0,a}$ instead of $\mathcal{G}^{(s,p)}_{\sigma,x_0,a}$.

For later purposes in the course of this chapter, in the following we will give certain results on commutators of the transformation operators $\mathcal{G}$ and $\mathcal{G}_\sigma$ with the Laplacian and the gradient.

Lemma 3.8. *1. We define*

$$[\Delta, \mathcal{G}^{-1}_{\sigma,x_0,a}] : W^{2,p}(\hat{Q}_{x_0,a} \cap \mathbb{R}^n_+)^n \to L^p(Q_{x_0,a} \cap \Omega)^n$$

with

$$[\Delta, \mathcal{G}^{-1}_{\sigma,x_0,a}]\hat{u} := \Delta\mathcal{G}^{-1}_{\sigma,x_0,a}\hat{u} - \mathcal{G}^{-1}_{\sigma,x_0,a}\Delta\hat{u}.$$

Then there exist constants $C > 0$ and $b_2 \in (0, b_1)$, such that for all $a \in (0, b_2)$ and $x_0 \in \partial\Omega$ the inequality

$$\|[\Delta, \mathcal{G}^{-1}_{\sigma,x_0,a}]\hat{u}\|_{L^p(Q_{x_0,a}\cap\Omega)^n} \leq C(1+\|h_{x_0,a}\|_{W^{3,\infty}(\mathbb{R}^{n-1})})^3\|\hat{u}\|_{W^{2,p}(\hat{Q}_{x_0,a}\cap\mathbb{R}^n_+)^n} \quad (3.8)$$

is valid for all $\hat{u} \in W^{2,p}(\hat{Q}_{x_0,a} \cap \mathbb{R}^n_+)^n$.

2. There exist constants $C > 0$ and $b_3 \in (0, b_2)$, such that for all $a \in (0, b_3)$ and $x_0 \in \partial\Omega$ the inequality

$$\|\left(\nabla\mathcal{G}^{-1}_{x_0,a} - \mathcal{G}^{-1}_{\sigma,x_0,a}\nabla\right)\hat{p}\|_{L^p(Q_{x_0,a}\cap\Omega)^n} \leq C\|\nabla h_{x_0,a}\|_\infty\|\nabla\hat{p}\|_{L^p(\hat{Q}_{x_0,a}\cap\mathbb{R}^n_+)^n} \quad (3.9)$$

holds for all $\hat{p} \in \widehat{W}^{1,p}(\hat{Q}_{x_0,a} \cap \mathbb{R}^n_+)$.

Proof. 1. Let $\hat{u} \in W^{2,p}(\hat{Q}_{x_0,a} \cap \mathbb{R}^n_+)^n$. Then we obtain

$$\begin{aligned} & \|\mathcal{G}^{-1}_{\sigma,x_0,a}\Delta\hat{u}\|_{L^p(Q_{x_0,a}\cap\Omega)^n} \\ \leq\ & \|\mathcal{G}^{-1}_{\sigma,x_0,a}\|_{\mathscr{L}(L^p(\hat{Q}_{x_0,a}\cap\mathbb{R}^n_+)^n, L^p(Q_{x_0,a}\cap\Omega)^n)}\|\Delta\hat{u}\|_{L^p(\hat{Q}_{x_0,a}\cap\mathbb{R}^n_+)^n} \\ \leq\ & (1+\|\nabla h_{x_0,a}\|_\infty)\|\hat{u}\|_{W^{2,p}(\hat{Q}_{x_0,a}\cap\mathbb{R}^n_+)^n}. \end{aligned}$$

The same estimates as in the proof of the first assertion of Lemma 3.6 yield

$$\begin{aligned} & \|\Delta\mathcal{G}^{-1}_{\sigma,x_0,a}\hat{u}\|_{L^p(Q_{x_0,a}\cap\Omega)^n} \\ \leq\ & \|\Delta\|_{\mathscr{L}(W^{2,p}(Q_{x_0,a}\cap\Omega)^n, L^p(Q_{x_0,a}\cap\Omega)^n)}\|\mathcal{G}^{-1}_{\sigma,x_0,a}\hat{u}\|_{W^{2,p}(Q_{x_0,a}\cap\Omega)^n} \\ \leq\ & C(1+\|h_{x_0,a}\|_{W^{3,\infty}(\mathbb{R}^{n-1})})^3\|\hat{u}\|_{W^{2,p}(\hat{Q}_{x_0,a}\cap\mathbb{R}^n_+)^n}. \end{aligned}$$

Together these estimates give the claim.

2. Let $\hat{p} \in \widehat{W}^{1,p}(\hat{Q}_{x_0,a} \cap \mathbb{R}^n_+)$. Then we have

$$\begin{aligned}
& \| \left(\nabla \mathcal{G}^{-1}_{x_0,a} - \mathcal{G}^{-1}_{\sigma,x_0,a} \nabla\right) \hat{p} \|_{L^p(Q_{x_0,a} \cap \Omega)^n} \\
= \; & \|\nabla \hat{p}(g_{x_0,a}(\cdot)) - J^{-1}_{g_{x_0,a}} (\nabla \hat{p})(g_{x_0,a}(\cdot))\|_{L^p(Q_{x_0,a} \cap \Omega)^n} \\
= \; & \|(\nabla \hat{p})(g_{x_0,a}(\cdot)) J^{-1}_{g_{x_0,a}} - J^{-1}_{g_{x_0,a}} (\nabla \hat{p})(g_{x_0,a}(\cdot))\|_{L^p(Q_{x_0,a} \cap \Omega)^n} \\
\leq \; & \|(\nabla \hat{p})(g_{x_0,a}(\cdot)) \nabla h_{x_0,a}\|_{L^p(Q_{x_0,a} \cap \Omega)^n} \\
& + \|(\partial_n \hat{p})(g_{x_0,a}(\cdot)) \nabla h_{x_0,a}\|_{L^p(Q_{x_0,a} \cap \Omega)^n} \\
\leq \; & C \|\nabla h_{x_0,a}\|_\infty \|\nabla \hat{p}\|_{L^p(\hat{Q}_{x_0,a} \cap \mathbb{R}^n_+)^n},
\end{aligned}$$

which proves the claim. □

Remark 3.9. In the foregoing constructions we assumed $x_0 \in \partial\Omega$. If we have $x_0 \in \Omega$, then all the transformations may be performed in the same way by choosing U, such that $U \cap \partial\Omega = \emptyset$, and $h = 0$.

In the following we will state some facts concerning certain mapping properties of Bogovskiĭ's operator. These will be necessary for the construction of a localization procedure, that respects the condition of solenoidality.

We remind the reader that for $g \in L^p(\Omega)$ with $\int_{\Omega'} g \, \mathrm{d}x = 0$ and a bounded Lipschitz domain $\Omega' \subset \Omega$ the problem

$$\begin{cases} \operatorname{div} u = g & \text{in } \Omega', \\ u = 0 & \text{on } \partial\Omega', \end{cases} \tag{3.10}$$

admits a solution $u \in W^{1,p}_0(\Omega')^n$, which is given by $u = B_{\Omega'} g$, where $B_{\Omega'}$ denotes the Bogovskiĭ operator on the domain Ω'. Moreover, for $s > -2 + \frac{1}{p}$ the inequality

$$\|B_{\Omega'} f\|_{W^{s+1,p}_0(\Omega')^n} \leq C \|f\|_{W^{s,p}_0(\Omega')} \tag{3.11}$$

holds for every $f \in W^{s,p}_0(\Omega')$ (see section 1.4).

Finally, we will prove a result giving an estimate of the operator norm of the Bogovskiĭ operator on anullis, which will be essential for later purposes.

Lemma 3.10. *For an $x \in \overline{\Omega}$ and some $r > 0$ we consider the set*

$$\Omega_{x,r} := (Q_{x,r} \setminus Q_{x,3r/4}) \cap \Omega \tag{3.12}$$

and let b_3 be given as in Lemma 3.8.

Then there exist constants $C > 0$ and $b \in (0, b_3)$, such that for all $a \in (0, b)$ and $x \in \partial\Omega$ or $x \in \Omega$ with $\overline{Q_{x,a}} \subset \Omega$

$$\|B_{\Omega_{x,a}} f\|_{L^p(\Omega_{x,a})^n} \leq C a \|f\|_{L^p(\Omega_{x,a})}, \quad f \in L^p(\Omega_{x,a}). \tag{3.13}$$

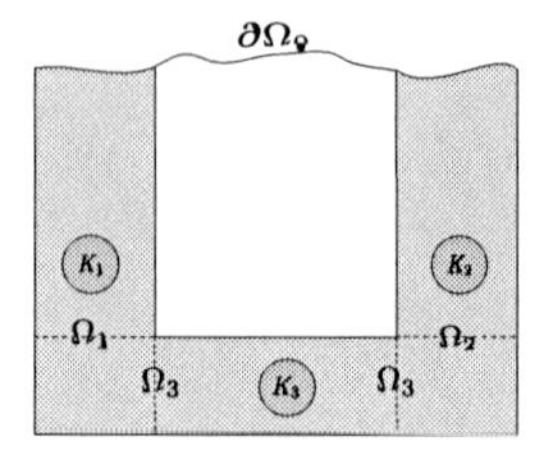

Figure 3.1: Bogovskiĭ's operator on anullis

Proof. First, by Poincaré's inequality we have

$$\|B_{\Omega_{x,a}} f\|_{L^p(\Omega_{x,a})^n} \leq Ca \|\nabla B_{\Omega_{x,a}} f\|_{L^p(\Omega_{x,a})^{n^2}}. \tag{3.14}$$

Now, consider the Bogovskiĭ operator on anullis Q_L with diameter one, where L denotes the Lipschitz constant of a domain Ω_0, cf. figure 3.1. In this setting Ω_3 is star-shaped with respect to any $x \in K_3$. Moreover, there exists $L_0 > 0$, such that Ω_1 and Ω_2 are star-shaped with respect to K_1 and K_2, respectively, for any $L < L_0$. Therefore, from [Gal94a, Lemma III.3.2 and Theorem III.3.1]

$$\|\nabla B_{Q_L} f\|_{L^p(Q_L)^{n^2}} \leq C\|f\|_{L^p(Q_L)}, \quad L \leq L_0, \; f \in L^p(Q_L) \tag{3.15}$$

follows. By inequality (3.7) we may choose a, such that the Lipschitz constant of $h_{x_0,a}$ is sufficiently small. Then, (3.14) and (3.15) together with a scaling transformation $t : \mathbb{R}^n \to \mathbb{R}^n$ defined by $t(x) := ax$ yield

$$\begin{aligned}
\|B_{\Omega_{x,a}} f\|_{L^p(\Omega_{x,a})^n} &\leq Ca \|\nabla B_{\Omega_{x,a}} f\|_{L^p(\Omega_{x,a})^{n^2}} \\
&= Ca^{n+1} \|\nabla B_{Q_L}(f \circ t)\|_{L^p(Q_L)^{n^2}} \\
&\leq Ca^{n+1} \|f \circ t\|_{L^p(Q_L)} \\
&= Ca \|f\|_{L^p(\Omega_{x,a})}.
\end{aligned}$$

Thus, the claim follows in the case $x \in \partial\Omega$.

For $x \in \Omega$ the corresponding domain $\Omega_{x,a}$ is a finite union of cuboids. Of course, these cuboids each are star-shaped with respect to a respective set. Hence, the assertion may be proved by applying the same arguments. □

3.3 $\mathcal{R}$-sectoriality of the Stokes operator in the case of bounded domains

The case of a bounded domain will be treated by a localization procedure. For this purpose, we start with the following covering lemma, which allows to decompose the original problem into several localized problems on a finite number of subsets of Ω.

Lemma 3.11. *Let $\Omega \subset \mathbb{R}^n$ be a bounded Lipschitz domain and b be given as in Lemma 3.10. Then there exist $c \in (0, b)$, $A_0 > 1$, and $N_0 > 0$, such that for all $a \in (0, c)$ there is a finite covering of Ω consisting of open cubes $\{Q(x_j, 3a_j/4)\}_{j=1}^N$ with center x_j and sidelength a_j, where either $a_j = a$ and $x_j \in \partial\Omega$ or $a_j = a/A_0$ and $\overline{Q(x_j, a_j)} \subset \Omega$, and at most N_0 different sets $Q(x_j, a_j)$ have nonempty intersection.*

Proof. Take the covering $\bigcup_{x\in\Omega} Q(x, 3a_x/4) \supset \overline{\Omega}$, where $a_x = a$, if $x_j \in \partial\Omega$, and $a_j = a/A_0$ for some $A_0 > 1$, if $x \in \Omega$. Since $\overline{\Omega} \subset \mathbb{R}^n$ is compact and has a Lipschitz boundary, we may choose A_0, such that there exists a finite subcovering $\{Q(x_j, 3a_j/4)\}_{j=1}^N$ satisfying $\overline{Q(x_j, a_j)} \subset \Omega$, if $x_j \in \Omega$. Now [SW71, Lemma II.3.3] yields the assertion. □

Now, we choose a partition of unity $\{\varphi_j\}_{j=1}^N$ subordinated to the covering $\{Q(x_j, 3a_j/4)\}_{j=1}^N$ of Ω. Furthermore, we choose cut-off functions $\psi_j \in C_c^\infty(Q(x_j, a_j))$, such that $0 \leq \psi_j \leq 1$ and $\psi_j \equiv 1$ on $\overline{Q(x_j, 3a_j/4)}$.

Fix $\theta \in (0, \pi)$ and let $\lambda \in \Sigma_\theta$. Recall from (3.12) that Ω_{x_j,a_j} is defined as

$$\Omega_{x_j,a_j} = (Q_{x_j,a_j} \setminus Q_{x_j,3a_j/4}) \cap \Omega. \tag{3.16}$$

For $f \in L^p_\sigma(\Omega)$ define

$$f_j := \psi_j f - B_{\Omega_{x_j,a_j}}((\nabla\psi_j)f)$$

and let $\hat{f}_j$ denote the extension to $\mathbb{R}^n_+$ by 0 of the push-forward $\mathcal{G}_{\sigma,x_j,a_j} f_j$ related to Q_{x_j,a_j}.

Since $\nabla\psi_j \equiv 0$ on $\overline{Q(x_j, 3a_j/4)}$, we obtain by integration by parts and the Gauß theorem

$$\begin{aligned}
\int_{\Omega_{x_j,a_j}} (\nabla\psi_j) f \, \mathrm{d}x &= \int_{Q_{x_j,a_j}} (\nabla\psi_j) f \, \mathrm{d}x \\
&= \int_{\partial Q_{x_j,a_j}} \psi_j(\nu \cdot f) \, \mathrm{d}\sigma(x) - \int_{Q_{x_j,a_j}} \psi_j \operatorname{div} f \, \mathrm{d}x \\
&= 0.
\end{aligned}$$

Hence, due to the properties of the Bogovskiĭ operator we have

$$\operatorname{div} f_j = \psi_j \operatorname{div} f + (\nabla\psi_j) f - \operatorname{div} B_{\Omega_{x_j,a_j}}((\nabla\psi_j)f) = 0.$$

Since $\|\nabla \psi_j\|_{L^\infty(Q(x_j,a_j))^n} \leq C/a$ for some constant $C > 0$, independent of a and j, assertion (3.13) yields $\hat{f}_j \in L^p_\sigma(\mathbb{R}^n_+)$ and

$$\|\hat{f}_j\|_{L^p(\mathbb{R}^n_+)^n} \leq C\|f\|_{L^p(Q_{x_j,a_j}\cap\Omega)^n}, \tag{3.17}$$

where $C > 0$ is independent of a, j, and f. Since $\hat{f}_j \in L^p_\sigma(\mathbb{R}^n_+)$, by the results of Section 2.1, there exists a unique solution $(\hat{u}_j, \hat{p}_j)$ to the equation

$$\begin{cases} \lambda\hat{u}_j - \Delta\hat{u}_j + \nabla\hat{p}_j &= \hat{f}_j \quad \text{in } \mathbb{R}^n_+, \\ \operatorname{div}\hat{u}_j &= 0 \quad \text{in } \mathbb{R}^n_+, \\ \hat{u}_j &= 0 \quad \text{on } \partial\mathbb{R}^n_+. \end{cases}$$

We set

$$\begin{aligned} u_\lambda &:= \tilde{R}(\lambda)f := \sum_{j=1}^N (\varphi_j u_j - B_\Omega((\nabla\varphi_j)u_j)), \\ p_\lambda &:= \sum_{j=1}^N \varphi_j p_j, \end{aligned}$$

where $u_j := \mathcal{G}^{-1}_{\sigma,x_j,a_j}(\hat{u}_j|_{\hat{Q}_{x_j,a_j}})$ and $p_j := \mathcal{G}^{-1}_{x_j,a_j}(\hat{p}_j|_{\hat{Q}_{x_j,a_j}})$. Every addend of $\tilde{R}(\lambda)$ as defined above is a composition of bounded operators and the resolvent of the negative Stokes operator in the halfspace. Since the sum is finite, it follows from (2.4) that

$$\mathcal{R}\{\lambda\tilde{R}(\lambda) : \lambda \in \Sigma_\theta\} \leq C, \tag{3.18}$$

where C depends on the covering. Moreover, due to Lemma 3.6 we have $(u_\lambda, p_\lambda) \in (W^{2,p}(\Omega)^n \cap W^{1,p}_0(\Omega)^n \cap L^p_\sigma(\Omega)) \times \widehat{W}^{1,p}(\Omega)$. Putting (u_λ, p_λ) into (3.3) yields

$$\begin{aligned} \lambda u_\lambda - \Delta u_\lambda + \nabla p_\lambda &= \sum_{j=1}^N (\lambda - \Delta)(\varphi_j u_j) - \sum_{j=1}^N (\lambda - \Delta) B_\Omega((\nabla\varphi_j)u_j) \\ &\quad + \sum_{j=1}^N (\nabla\varphi_j)p_j + \sum_{j=1}^N \varphi_j \nabla p_j. \end{aligned}$$

Furthermore, setting $[\Delta, \varphi_j]u_j := \Delta(\varphi_j u_j) - \varphi_j \Delta u_j$ we have

$$\sum_{j=1}^N (\lambda - \Delta)(\varphi_j u_j) = \sum_{j=1}^N \varphi_j(\lambda - \Delta)u_j - [\Delta, \varphi_j]u_j.$$

Finally, we calculate

$$\begin{aligned}\sum_{j=1}^{N}\varphi_j(\lambda-\Delta)u_j &= \sum_{j=1}^{N}\varphi_j\mathcal{G}^{-1}_{\sigma,x_j,a_j}(\lambda-\Delta)\hat{u}_j-\varphi_j[\Delta,\mathcal{G}^{-1}_{\sigma,x_j,a_j}]\hat{u}_j\\ &= \sum_{j=1}^{N}\varphi_j\mathcal{G}^{-1}_{\sigma,x_j,a_j}(\hat{f}_j-\nabla\hat{p}_j)-\varphi_j[\Delta,\mathcal{G}^{-1}_{\sigma,x_j,a_j}]\hat{u}_j\\ &= f-\sum_{j=1}^{N}\varphi_j\mathcal{G}^{-1}_{\sigma,x_j,a_j}\nabla\hat{p}_j-\varphi_j[\Delta,\mathcal{G}^{-1}_{\sigma,x_j,a_j}]\hat{u}_j.\end{aligned}$$

Collecting the terms calculated above, we finally obtain that (u_λ, p_λ) satisfies

$$\left\{\begin{array}{rcll}\lambda u_\lambda-\Delta u_\lambda+\nabla p_\lambda &=& f+T_\lambda f & \text{in } \Omega,\\ \operatorname{div} u_\lambda &=& 0 & \text{in } \Omega,\\ u_\lambda &=& 0 & \text{on } \partial\Omega,\end{array}\right.$$

where the error term T_λ is given by

$$\begin{aligned}T_\lambda f &= \sum_{j=1}^{N}\varphi_j(\nabla\mathcal{G}^{-1}_{x_j,a_j}\hat{p}_j-\mathcal{G}^{-1}_{\sigma,x_j,a_j}\nabla\hat{p}_j)-\sum_{j=1}^{N}(\lambda-\Delta)B_\Omega((\nabla\varphi_j)u_j)\\ &\quad -\sum_{j=1}^{N}\Big([\Delta,\varphi_j]u_j+\varphi_j[\Delta,\mathcal{G}^{-1}_{\sigma,x_j,a_j}]\hat{u}_j\Big)+\sum_{j=1}^{N}(\nabla\varphi_j)p_j\\ &=: T^1_\lambda f+T^2_\lambda f+T^3_\lambda f+T^4_\lambda f.\end{aligned}$$

Lemma 3.12. *Let $\theta\in(0,\pi)$, $\lambda_0>1$, and $\alpha\in(0,\frac{1}{2p'})$. If a is chosen sufficiently small in the foregoing segmentation of Ω, then there exist $C>0$ and $s\in(\frac{1}{p},1)$, independent of λ_0 and α, such that*

1. $\mathcal{R}_{L^p_\sigma(\Omega)}\{PT^1_\lambda:\lambda\in\lambda_0+\Sigma_\theta\}\leq\frac{1}{3}$,
2. $\mathcal{R}_{L^p_\sigma(\Omega)}\{PT^2_\lambda:\lambda\in\lambda_0+\Sigma_\theta\}\leq\frac{C}{\lambda_0^{(1-s)\alpha}}$,
3. $\mathcal{R}_{L^p_\sigma(\Omega)}\{PT^3_\lambda:\lambda\in\lambda_0+\Sigma_\theta\}\leq\frac{C}{\sqrt{\lambda_0}}+\frac{1}{3}$,
4. $\mathcal{R}_{L^p_\sigma(\Omega)}\{PT^4_\lambda:\lambda\in\lambda_0+\Sigma_\theta\}\leq\frac{C}{\lambda_0^\alpha}$.

In particular, by the calculation rules for $\mathcal{R}$-bounds these estimates imply that $\{PT_\lambda:\lambda\in\lambda_0+\Sigma_\theta\}\subset\mathscr{L}(L^p_\sigma(\Omega))$ is an $\mathcal{R}$-bounded family of operators with $\mathcal{R}$-bound

$$\mathcal{R}_{L^p_\sigma(\Omega)}\{PT_\lambda:\lambda\in\lambda_0+\Sigma_\theta\}\leq\frac{2}{3}+\frac{C_1}{\lambda_0}+\frac{C_2}{\lambda_0^{(1-s)\alpha}}+\frac{C_3}{\lambda_0^\alpha}.$$

Proof. First, we prove the estimates of T_λ^1, T_λ^2, T_λ^3, and T_λ^4 separately. By $A_{\mathbb{R}^n_+}$ we denote the Stokes operator in $L^p_\sigma(\mathbb{R}^n_+)$.

1. We may write $\nabla \hat{p}_j = (\mathrm{Id} - P)\Delta(\lambda + A_{\mathbb{R}^n_+})^{-1}\hat{f}_j$. Furthermore, by Lemma 3.11 at most N_0 different sets $Q(x_j, a_j)$ have nonempty intersection. Therefore the estimates (2.8), (3.9), and (3.17) yield

$$\mathcal{R}_{L^p_\sigma(\Omega)}\{PT_\lambda^1 : \lambda \in \lambda_0 + \Sigma_\theta\} \le C(1 + \|P\|) \max_{1\le j\le N}\{\|\nabla h_{x_j,a_j}\|_\infty\}.$$

Hence, by (3.7), we may choose $b_1 > 0$ small enough, such that

$$\mathcal{R}_{L^p_\sigma(\Omega)}\{PT_\lambda^1 : \lambda \in \lambda_0 + \Sigma_\theta\} \le \frac{1}{3}, \quad a \in (0, b_1).$$

2. We split the term $T_\lambda^2 f$ as follows:

$$\begin{aligned}
-\sum_{j=1}^N (\lambda - \Delta)B_\Omega((\nabla\varphi_j)u_j) \;=\; & -\sum_{j=1}^N B_\Omega((\nabla\varphi_j)f_j) + \sum_{j=1}^N \Delta B_\Omega((\nabla\varphi_j)u_j) \\
& -\sum_{j=1}^N B_\Omega((\nabla\varphi_j)\mathcal{G}^{-1}_{\sigma,x_j,a_j}(\Delta\hat{u}_j|_{\hat{Q}_{x_j,a_j}})) \\
& +\sum_{j=1}^N B_\Omega((\nabla\varphi_j)\mathcal{G}^{-1}_{\sigma,x_j,a_j}(\nabla\hat{p}_j|_{\hat{Q}_{x_j,a_j}})).
\end{aligned}$$

Since $\operatorname{supp} B_{\Omega_{x_j,a_j}}((\nabla\psi_j)f) \subset \Omega_{x_j,a_j}$, the first component of T_λ^2 disappears due to the properties of the cut-off functions φ_j and ψ_j as the calculation

$$\begin{aligned}
\sum_{j=1}^N B_\Omega((\nabla\varphi_j)f_j) &= \sum_{j=1}^N B_\Omega((\nabla\varphi_j)\psi_j f) - B_\Omega\big((\nabla\varphi_j)B_{\Omega_{x_j,a_j}}((\nabla\psi_j)f\big) \\
&= B_\Omega(\nabla\sum_{j=1}^N \varphi_j)f) \\
&= 0
\end{aligned}$$

shows.

For the treatment of the second and the third component, we recall that u_j is given by $u_j = \mathcal{G}^{-1}_{\sigma,x_j,a_j}(\hat{u}_j|_{\hat{Q}_{x_j,a_j}})$ and that we have $\hat{u}_j = (\lambda + A_{\mathbb{R}^n_+})^{-1}\hat{f}_j$.

If we take $s = 1$ in Lemma 2.3, we obtain by estimate (3.11) and the boundedness of $\mathcal{G}^{-1}_{\sigma,x_j,a_j}$ that the $\mathcal{R}$-bound of the second component of T_λ^2 is less than $C\lambda_0^{-\frac{1}{2}}$.

In order to estimate the third one, fix $r \in (\frac{1}{p}, 1)$. By Lemma 2.3, there exists $C > 0$, such that

$$\mathcal{R}_{L^p_\sigma(\mathbb{R}^n_+)\to W^{-1+r,p}(\mathbb{R}^n_+)^n}\{\Delta(\lambda + A_{\mathbb{R}^n_+})^{-1} : \lambda \in \lambda_0 + \Sigma_\theta\} \le C\lambda_0^{-\frac{1-r}{2}}.$$

By Proposition 1.6 we have $W_0^{-1+r,p}(\mathbb{R}^n_+) = W^{-1+r,p}(\mathbb{R}^n_+)$ for our choice of r. Since supp $\nabla\varphi_j \subset Q_{x_j,a_j}$, this assures that $(\nabla\varphi_j)\mathcal{G}^{-1}_{\sigma,x_j,a_j}(\Delta\hat{u}_j|_{\hat{Q}_{x_j,a_j}}) \in W_0^{-1+r,p}(\Omega)$. Thus, it follows from (3.11) again that the $\mathcal{R}$-bound of the third component of T^2_λ is less than $C\lambda_0^{-\frac{1-r}{2}}$.

Similarly, by Lemma 2.6, where we take $s = r$, and (3.11), the $\mathcal{R}$-bound of the last component is less than $\lambda_0^{-(1-s)\alpha}$ provided $\alpha \in (0, \frac{1}{2p'})$.

Summarizing, there exists $C > 0$, such that

$$\lambda_0^{(1-s)\alpha}\mathcal{R}_{L^p_\sigma(\Omega)}\{PT^2_\lambda : \lambda \in \lambda_0 + \Sigma_\theta\} \le C, \quad a \in (0, b_2), \lambda_0 \ge 1.$$

3. First note that $[\Delta, \varphi_j]u_j = (\Delta\varphi_j)u_j + 2\nabla\varphi_j \cdot \nabla u_j$. Thus, $[\Delta, \varphi_j]$ is a first order differential operator. Moreover, for the term $[\Delta, \mathcal{G}^{-1}_{\sigma,x_j,a_j}]\hat{u}_j$ estimate (3.8) holds. As above, we will use that $u_j = \mathcal{G}^{-1}_{\sigma,x_j,a_j}(\hat{u}_j|_{\hat{Q}_{x_j,a_j}})$ and $\hat{u}_j = (\lambda + A_{\mathbb{R}^n_+})^{-1}\hat{f}_j$. Hence, applying Lemma 2.3 with $s = 1$ to the first term of T^3_λ as well as estimates (2.8) and (3.8) to the second one, we obtain that there exist $b_2 \in (0, b_1)$ and $C > 0$, such that

$$\mathcal{R}_{L^p_\sigma(\Omega)}\{PT^3_\lambda : \lambda \in \lambda_0 + \Sigma_\theta\} \le \frac{C}{\sqrt{\lambda_0}} + \frac{1}{3}, \quad a \in (0, b_2).$$

4. Choosing $s = 0$ in Lemma 2.6, we conclude that for $\alpha \in (0, \frac{1}{2p'})$ there exists $C > 0$, such that

$$\lambda_0^\alpha\mathcal{R}_{L^p_\sigma(\Omega)}\{PT^4_\lambda : \lambda \in \lambda_0 + \Sigma_\theta\} \le C, \quad a \in (0, b_2), \lambda_0 \ge 1.$$

Combining these estimates, in the segmentation of Ω we may choose a small enough, such that $\{PT_\lambda : \lambda \in \lambda_0 + \Sigma_\theta\} \subset \mathscr{L}(L^p_\sigma(\Omega))$ is $\mathcal{R}$-bounded with the given $\mathcal{R}$-bound. This completes the proof of Lemma 3.12. □

Having the foregoing Lemma at hands, we are able to prove the main result of this chapter in the case of a bounded domain Ω.

Proof of Theorem 3.1 for bounded domains.

We fix $\theta \in (0, \pi)$ and take a segmentation of Ω with sufficiently small a. Then, by Lemma 3.12 there exist $\lambda_0 > 0$, such that

$$\mathcal{R}_{L^p_\sigma(\Omega)}\{PT_\lambda : \lambda \in \lambda_0 + \Sigma_\theta\} < 1. \tag{3.19}$$

Therefore, the mapping $R(\lambda) : L^p_\sigma(\Omega) \to D(A)$, defined by

$$R(\lambda) := \tilde{R}(\lambda)(\mathrm{Id} + PT_\lambda)^{-1}$$

is well-defined via the Neumann series and we have

$$(\lambda + A)R(\lambda)f = f, \quad f \in L^p_\sigma(\Omega), \lambda \in \lambda_0 + \Sigma_\theta. \tag{3.20}$$

Moreover, by (3.18) and (3.19), there exists a constant $C > 0$, such that

$$\mathcal{R}_{L^p_\sigma(\Omega)}\{\lambda R(\lambda) : \lambda \in \lambda_0 + \Sigma_\theta\} \leq C. \tag{3.21}$$

Now, consider the Stokes operator A_2 in $L^2_\sigma(\Omega)$. An integration by parts yields

$$\langle -A_2 u, u\rangle = -\|\nabla u\|_2^2 \leq 0, \quad u \in L^2_\sigma(\Omega),$$

which means that $-A_2$ is dissipative. Therefore, $\lambda_0+\Sigma_\theta \subset \rho(A_2)$ and $R_2(\lambda) = (\lambda + A_2)^{-1}$ for $\lambda \in \lambda_0 + \Sigma_\theta$. Hence, by (3.21)

$$\mathcal{R}_{L^2_\sigma(\Omega)}\{\lambda(\lambda + A_2)^{-1} : \lambda \in \lambda_0 + \Sigma_\theta\} \leq C.$$

Now, let $2 \leq p < \infty$. Since we consider a bounded domain Ω, we have $L^p(\Omega)^n \subset L^2(\Omega)^n$ and $D(A_p) \subset D(A_2)$. Thus, $\lambda + A_p$ is injective for $\lambda \in \lambda_0 + \Sigma_\theta$. Equation (3.20) implies that $\lambda + A_p$ is also surjective. It follows that $R(\lambda) = (\lambda + A_p)^{-1}$ and we obtain from (3.21) that

$$\mathcal{R}_{L^p_\sigma(\Omega)}\{\lambda(\lambda + A_p)^{-1} : \lambda \in \lambda_0 + \Sigma_\theta\} \leq C \tag{3.22}$$

for $2 \leq p < \infty$. In particular, $\lambda_0 + A_p$ is a sectorial operator. Hence, $-A_p$ is the generator of an analytic semigroup on $L^p_\sigma(\Omega)$.

Finally, let $1 < p \leq 2$ and $1/p + 1/p' = 1$. Then the adjoint operator $-A'_{p'}$ generates an analytic semigroup on $L^p_\sigma(\Omega)$ as follows from Remark 1.15. Integration by parts yields $A_p \subset A'_{p'}$. Since $\lambda + A_p$ is surjective, A_p and $A'_{p'}$ coincide. Hence, inequality (3.22) holds for $1 < p \leq 2$, too.

Thus, it remains to show that we may choose $\lambda_0 < 0$. To this end, note that for $\varepsilon > 0$

$$\langle -A_2 u, u\rangle + \varepsilon\langle u, u\rangle \leq -\|\nabla u\|_2^2 + C\varepsilon\|\nabla u\|_2^2$$

holds thanks to Poincaré's inequality. That means that for ε small enough $-A_2 + \varepsilon\mathrm{Id} : D(A_2) \to L^2_\sigma(\Omega)$ is dissipative. Thus, $\|e^{-tA_2}\|_{\mathscr{L}(L^2(\Omega))} \leq e^{-\varepsilon t}$ for all $t \geq 0$. In particular, this yields $0 \in \rho(A_2)$. Since $W^{1,p}(\Omega)$ is compactly embedded in $L^p(\Omega)$ thanks to Rellich's embedding theorem and $D(A_p) \subset W^{1,p}(\Omega)$, the resolvent $(\lambda + A_p)^{-1}$ is compact for any $\lambda \in \Sigma_\theta$ and $1 < p < \infty$. It thus follows by [Are94, Prop. 2.6] that $\sigma(A_p) = \sigma(A_2)$ for $1 < p < \infty$. Since $-A_p$ generates an analytic semigroup, $\|e^{-tA_p}\|_{\mathscr{L}(L^p(\Omega))} \leq Ce^{-\varepsilon t}$ for all $t \geq 0$, see Remark 1.15. Therefore, by Proposition 1.32 A_p admits maximal L^p-regularity on $[0, \infty)$. Furthermore, since the growth bound of e^{-tA_p} is negative we can choose $\lambda_0 < 0$. The proof of Theorem 3.1 is complete in case of a bounded domain Ω.

□

3.4 Proof of the main result for exterior domains

In this section we will transfer the result from bounded domains to exterior domains Ω with C^3-boundary, i.e. $\Omega := \mathbb{R}^n \setminus K$ for some compact set K. The proof will be carried out by using the result on bounded domains together with the according result on $\mathbb{R}^n$. The procedure of combining the results from bounded domains and the whole space follows the method described in [GHH06a].

We will start with segmenting the exterior domain Ω into several subsets. For this purpose, we choose $R > 0$, such that $\Omega^c \subset B_R(0) := \{x \in \mathbb{R}^n : |x| < R\}$ and set

$$\begin{aligned} D &:= \Omega \cap B_{R+5}(0), \\ K_1 &:= \{x \in \Omega : R < |x| < R+3\}, \\ K_2 &:= \{x \in \Omega : R+2 < |x| < R+5\}. \end{aligned}$$

Furthermore, we choose cut-off functions $\varphi, \psi \in C^\infty(\mathbb{R}^n)$, such that $0 \leq \varphi, \psi \leq 1$ and

$$\varphi(x) = \begin{cases} 0, & |x| \leq R+1, \\ 1, & |x| \geq R+2, \end{cases} \qquad \text{and} \qquad \psi(x) = \begin{cases} 1, & |x| \leq R+3, \\ 0, & |x| \geq R+4. \end{cases}$$

Our aim is to formulate system (3.3) in terms of a system in a bounded domain and in $\mathbb{R}^n$, respectively. To this end, for $f \in L^p_\sigma(\Omega)$ we define the extension f^R to all of $\mathbb{R}^n$ by

$$f^R(x) = \begin{cases} f(x), & x \in \Omega, \\ 0, & \text{else}. \end{cases}$$

Since $C^\infty_{c,\sigma}(\Omega)$ is dense in $L^p_\sigma(\Omega)$, we have that f^R belongs to $L^p_\sigma(\mathbb{R}^n)$. On the other hand, we set $f^D := \psi f - B_{K_2}((\nabla\psi)f)$. Since $\int_{K_2} (\nabla\psi) f \,\mathrm{d}x = \int_{K_2} \operatorname{div}(\psi f)\,\mathrm{d}x = 0$ in view of the Gauß theorem, it follows that $f^D \in L^p_\sigma(D)$.

For $\lambda \in \Sigma_\theta$ for some $0 < \theta < \pi$ consider the solutions u^R_λ and $(u^D_\lambda, p^D_\lambda)$ to the equations

$$\begin{cases} \lambda u^R_\lambda - \Delta u^R_\lambda &= f^R \quad \text{in } \mathbb{R}^n, \\ \operatorname{div} u^R_\lambda &= 0 \quad \text{in } \mathbb{R}^n, \end{cases}$$

and

$$\begin{cases} \lambda u^D_\lambda - \Delta u^D_\lambda + \nabla p^D_\lambda &= f^D \quad \text{in } D, \\ \operatorname{div} u^D_\lambda &= 0 \quad \text{in } D, \\ u^D_\lambda &= 0 \quad \text{on } \partial D, \end{cases}$$

respectively. Restricting u_λ^R to Ω and extending u_λ^D as well as p_λ^D by 0 to all of Ω we may define the mapping $\tilde{R}(\lambda)$ by

$$\tilde{R}(\lambda)f := \varphi u_\lambda^R + (1-\varphi)u_\lambda^D + B_{K_1}(\nabla\varphi(u_\lambda^R - u_\lambda^D)). \tag{3.23}$$

Furthermore, set $u := \tilde{R}(\lambda)f$ and $p := (1-\varphi)p_\lambda^D$. Then putting (u,p) into (3.3) we obtain

$$\begin{aligned} \lambda u - \Delta u + \nabla p &= -2(\nabla\varphi)(\nabla(u_\lambda^R - u_\lambda^D)) - (\Delta\varphi)(u_\lambda^R - u_\lambda^D) \\ &\quad + (\lambda - \Delta)B_{K_1}(\nabla\varphi(u_\lambda^R - u_\lambda^D)) - (\nabla\varphi)p_\lambda^D \\ &\quad + (1-\varphi)f^D + \varphi f^R. \end{aligned}$$

Since $(1-\varphi)f^D + \varphi f^R = f$, due to the choice of the cut-off functions φ and ψ this results in

$$\left\{ \begin{aligned} \lambda u - \Delta u + \nabla p &= f + T_\lambda f, && \text{in } \Omega, \\ \operatorname{div} u &= 0, && \text{in } \Omega, \\ u &= 0, && \text{on } \partial\Omega. \end{aligned} \right.$$

Here, the error term $T_\lambda f$ is given by

$$\begin{aligned} T_\lambda f &= -2(\nabla\varphi)(\nabla(u_\lambda^R - u_\lambda^D)) - (\Delta\varphi)(u_\lambda^R - u_\lambda^D) \\ &\quad + (\lambda - \Delta)B_{K_1}(\nabla\varphi(u_\lambda^R - u_\lambda^D)) - (\nabla\varphi)p_\lambda^D \\ &=: T_\lambda^1 f + T_\lambda^2 f + T_\lambda^3 f + T_\lambda^4 f. \end{aligned}$$

Note that we may write

$$u_\lambda^R - u_\lambda^D = (\lambda - \Delta_{\mathbb{R}^n})^{-1} f^R - (\lambda + A_D)^{-1} f^D,$$

where $\Delta_{\mathbb{R}^n}$ denotes the Laplacian in $L^p(\mathbb{R}^n)^n$ and A_D is the Stokes operator in $L^p_\sigma(D)$. We have already proved in the previous result for bounded domains that

$$\mathcal{R}_{L^p_\sigma(D)}\{(\lambda + A_D)^{-1} : \lambda \in \lambda_0 + \Sigma_\theta\} \le \frac{C}{\lambda_0}, \quad \lambda_0 > 0.$$

Moreover, by [DHP03, Theorem 5.5] we know that for $1 < p < \infty$ the negative Laplacian $-\Delta_{\mathbb{R}^n}$ is an $\mathcal{R}$-sectorial operator of angle 0 in $L^p(\mathbb{R}^n)$. Furthermore, in $L^p(\mathbb{R}^n)^n$ the Helmholtz projection P and the resolvent of the Laplacian commute. Hence, we conclude

$$\mathcal{R}_{L^p_\sigma(\mathbb{R}^n)}\{(\lambda - \Delta_{\mathbb{R}^n})^{-1} : \lambda \in \lambda_0 + \Sigma_\theta\} \le \frac{C}{\lambda_0}, \quad \lambda_0 > 0.$$

Hence, for the operator $\tilde{T}_\lambda$ defined by $\tilde{T}_\lambda f := u_\lambda^R - u_\lambda^D$ we have

$$\mathcal{R}_{L^p_\sigma(\Omega)}\{\tilde{T}_\lambda : \lambda \in \lambda_0 + \Sigma_\theta\} \le \frac{C}{\lambda_0}, \quad \lambda_0 > 0. \tag{3.24}$$

In particular, the estimate

$$\mathcal{R}_{L^p_\sigma(\Omega)}\{\lambda \tilde{R}(\lambda) : \lambda \in \lambda_0 + \Sigma_\theta\} \leq C, \quad \lambda_0 > 0, \tag{3.25}$$

follows, too.

Moreover, we have

$$\mathcal{R}_{L^p(\mathbb{R}^n) \to W^{2,p}(\mathbb{R}^n)}\{(\lambda - \Delta_{\mathbb{R}^n})^{-1} : \lambda \in \lambda_0 + \Sigma_\theta\} \leq C, \quad \lambda_0 > 0$$

(see for example [DHP01]). Together with our result for the bounded domain case, this implies

$$\mathcal{R}_{L^p(\mathbb{R}^n) \to W^{2,p}(\mathbb{R}^n)}\{\tilde{T}_\lambda : \lambda \in \lambda_0 + \Sigma_\theta\} \leq C, \quad \lambda_0 > 0. \tag{3.26}$$

Lemma 3.13. *Let $\theta \in (0, \pi)$. Then there exist $\lambda_0 > 0$, such that*

$$\mathcal{R}_{L^p_\sigma(\Omega)}\{PT_\lambda : \lambda \in \lambda_0 + \Sigma_\theta\} < 1. \tag{3.27}$$

Proof. First, we may estimate the $\mathcal{R}$-bound of each component of T_λ separately.

1. Due to Remark 2.4 we may apply Lemma 2.3 with $s = 1$ to obtain

$$\mathcal{R}_{L^p_\sigma(\Omega)}\{PT^1_\lambda : \lambda \in \lambda_0 + \Sigma_\theta\} \leq \frac{C}{\sqrt{\lambda_0}}, \quad \lambda_0 \geq 1.$$

2. From (3.24) directly follows that

$$\mathcal{R}_{L^p_\sigma(\Omega)}\{PT^2_\lambda : \lambda \in \lambda_0 + \Sigma_\theta\} \leq \frac{C}{\lambda_0}, \quad \lambda_0 > 0.$$

3. Applying Lemma 2.3 with $s = 1$ to (3.24) and (3.26), we obtain that

$$\mathcal{R}^n_{L^p_\sigma(\Omega) \to W^{1,p}(\Omega)}\{\tilde{T}_\lambda : \lambda \in \lambda_0 + \Sigma_\theta\} \leq \frac{C}{\sqrt{\lambda_0}}, \quad \lambda_0 \geq 1.$$

Since supp $\nabla\varphi \subset K_1$, we have $(\nabla\varphi)(u^R_\lambda - u^D_\lambda) \in W^{1,p}_0(K_1)$. Now, inequality (3.11) yields

$$\mathcal{R}_{L^p_\sigma(\Omega)}\{PT^3_\lambda : \lambda \in \lambda_0 + \Sigma_\theta\} \leq \frac{C}{\sqrt{\lambda_0}}, \quad \lambda_0 \geq 1$$

4. Choosing $s = 0$ in Lemma 2.6, we conclude that for $\alpha \subset (0, \frac{1}{2p'})$ there exists $C > 0$, such that Thanks to Remark 2.7 we apply Lemma 2.6 with $s = 0$ to obtain that for $\alpha \in (0, \frac{1}{2p'})$ there exists $C > 0$, such that

$$\mathcal{R}_{L^p_\sigma(\Omega)}\{PT^4_\lambda : \lambda \in \lambda_0 + \Sigma_\theta\} \leq \frac{C}{\lambda_0^\alpha}, \quad \lambda_0 \geq 1.$$

Now, it follows from the separate estimates of the single components of T_λ that we may choose $\lambda_0 > 0$, such that (3.27) holds, which proves the assertion. □

Having Lemma 3.13 at hands, via the Neumann series we may now define the operator $R(\lambda) : L^p_\sigma(\Omega) \to D(A)$ by

$$R(\lambda) := \tilde{R}(\lambda)(\mathrm{Id} + PT_\lambda)^{-1}.$$

By definition, this operator satisfies

$$(\lambda + A)R(\lambda)f = f, \quad f \in L^p_\sigma(\Omega), \lambda \in \lambda_0 + \Sigma_\theta. \tag{3.28}$$

Moreover, by (3.24) and Lemma 3.13, there exists a constant $C > 0$, such that

$$\mathcal{R}_{L^p_\sigma(\Omega)}\{\lambda R(\lambda) : \lambda \in \lambda_0 + \Sigma_\theta\} \leq C. \tag{3.29}$$

Let $1 < p < \infty$. Then due to (3.28) the operator $\lambda + A_p$ is surjective for $\lambda \in \lambda_0 + \Sigma_\theta$. Moreover, since $\overline{\lambda} + A_{p'} \subset (\lambda + A_p)'$, we have that $(\lambda + A_p)'$ is surjective, too. Hence, $\lambda + A$ is injective for all $1 < p < \infty$. That means that $R(\lambda) = (\lambda + A)^{-1}$ and we have

$$\mathcal{R}_{L^p_\sigma(\Omega)}\{\lambda(\lambda + A)^{-1} : \lambda \in \lambda_0 + \Sigma_\theta\} \leq C,$$

which completes the proof of Theorem 3.1.

Chapter 4

Weak solutions of perturbed Navier-Stokes equations

In this chapter we will state an existence result for weak L^2-solutions of perturbed Navier-Stokes equations, which holds in certain unbounded subdomains of $\mathbb{R}^3$. In particular, we consider $\Omega \subset \mathbb{R}^3$ to be a halfspace or an infinite layer. In these domains we consider the system of equations

$$\left\{ \begin{array}{rcll} \partial_t u - \nu\Delta u + \hat{B}u + (u \cdot \nabla)u + \nabla p &=& 0, & t > 0,\ x \in \Omega, \\ \operatorname{div} u &=& 0, & t > 0,\ x \in \Omega, \\ u(t,x) &=& 0, & t > 0,\ x \in \partial\Omega, \\ u(0,x) &=& u_0, & x \in \Omega, \end{array} \right. \tag{4.1}$$

where $\hat{B}$ denotes a linear operator.

Now, we consider $B = P\hat{B}$, where P denotes the Helmholtz projection. Then we may rewrite system (4.1) as an evolution equation in $L^p_\sigma(\Omega)$ of the form

$$\left\{ \begin{array}{rcll} u'(t) + (\nu A + B)u(t) + P(u(t) \cdot \nabla)u(t) &=& 0, & t > 0, \\ u(0) &=& u_0. & \end{array} \right. \tag{4.2}$$

Suppose now that $D(A^{\frac{1}{2}}) \subset D(B)$, where A is the Stokes operator in $L^2_\sigma(\Omega)$. Then the following definition is meaningful.

Definition 4.1. Let $u_0 \in L^2_\sigma(\Omega)$. We call $u : [0,\infty) \to L^2_\sigma(\Omega)$ a *weak solution* of equation (4.2), if for all $T > 0$

i) $u \in L^\infty((0,T), L^2_\sigma(\Omega)) \cap L^2((0,T), D(A^{\frac{1}{2}}))$ and

ii)
$$-\int_0^T \langle u(t), \phi\rangle h'(t)\, \mathrm{d}t + \nu \int_0^T \langle \nabla u(t), \nabla\phi\rangle h(t)\, \mathrm{d}t$$
$$+ \int_0^T \langle \hat{B}u(t), \phi\rangle h(t)\, \mathrm{d}t + \int_0^T \langle (u(t)\cdot\nabla)u(t), \phi\rangle h(t)\, \mathrm{d}t \quad = \quad \langle u_0, \phi\rangle h(0),$$

holds for all $\phi \in D(A^{\frac{1}{2}})$ and all $h \in C^1([0,T],\mathbb{R})$ with $h(T) = 0$.

Our approach is inspired by the methods developed by Miyakawa and Sohr [MS88] in order to construct weak L^2-solutions to the pure Navier-Stokes equations on exterior domains. We show that under certain conditions on the perturbing operator $\hat{B}$ we can transfer their existence result to the system (4.1). Although the assertion of our result is stated completely within the L^2-framework, our proof makes use of maximal regularity estimates for the perturbed Stokes operator for $p \neq 2$.

4.1 Existence of weak solutions

Now we state the main result of this chapter concerning weak solutions of (4.2) in $L^2_\sigma(\Omega)$.

Theorem 4.2. *Let $\Omega \subset \mathbb{R}^3$ be a halfspace, an infinite layer, or the whole of $\mathbb{R}^3$. For $1 < p < \infty$ let A_p denote the Stokes operator in $L^p_\sigma(\Omega)$ and consider B to be a closed and linear operator in $L^p_\sigma(\Omega)$, such that the following conditions hold:*

1. *$D(A_p) \subset D(B)$ and for $u \in D(A_2)$*

$$\|Bu\|_2 \leq C(\|u\|_2 + \|\nabla u\|_2) \tag{4.3}$$

holds;

2. *for every $1 < p < \infty$ there exists $r > 0$, such that the operator $r + (\nu A_p + B)$ admits maximal regularity in $L^p_\sigma(\Omega)$;*

3. *for $u \in H^1_0(\Omega)^3 \cap L^2_\sigma(\Omega)$ the estimate*

$$0 \leq \operatorname{Re}\langle (\nu A_2 + B)u, u\rangle \tag{4.4}$$

holds.

Then, for every initial value $u_0 \in L^2_\sigma(\Omega)$ there exists a weak solution to system (4.2).

Remark 4.3. The operator $\nu A+B$ is closed in $L^p_\sigma(\Omega)$ for all $1<p<\infty$, as we assume that it has maximal regularity. Having in mind Proposition 1.11, condition (4.4) implies that $-(\nu A+B)$ is a dissipative operator in $L^2_\sigma(\Omega)$. Due to Proposition 1.22 and Remark 1.29 we have $\nu A+B\in \mathrm{BIP}(L^2_\sigma(\Omega))$. Hence, the square root $(\nu A+B)^{\frac{1}{2}}$ is well defined in $L^2_\sigma(\Omega)$ and its domain is given by $D((\nu A+B)^{\frac{1}{2}})=H^1_0(\Omega)^3\cap L^2_\sigma(\Omega)$.

As a further ingredient of the proof we will make use of the following simple interpolation lemma for L^p-spaces.

Lemma 4.4. *Let $\Omega\subset\mathbb{R}^n$ be an open set. If $1<p<q<r<\infty$, then $L^p(\Omega)\cap L^r(\Omega)\subset L^q(\Omega)$ and*

$$\|u\|_q\le \|u\|_p^\alpha\|u\|_r^{1-\alpha},\qquad \textit{where } \alpha=\frac{q^{-1}-r^{-1}}{p^{-1}-r^{-1}}.$$

Proof. For $\alpha=\frac{q^{-1}-r^{-1}}{p^{-1}-r^{-1}}$ we define $s:=\frac{p}{\alpha q}$, $t:=\frac{r}{(1-\alpha)q}$. Then s and t are conjugated Hölder exponents. Therefore we obtain by Hölder's inequality

$$\begin{aligned}\int_\Omega |u|^q\,\mathrm{d}x &\le \left(\int_\Omega |u|^{\alpha q s}\,\mathrm{d}x\right)^{\frac{1}{s}}\left(\int_\Omega |u|^{(1-\alpha q)t}\,\mathrm{d}x\right)^{\frac{1}{t}}\\ &= \|u\|_p^{\alpha q}\|u\|_r^{(1-\alpha)q}.\end{aligned}$$

Exponentiating with $\frac{1}{q}$ yields the claim. □

Proof of Theorem 4.2.

In the following, for simplicity of notation we will write $A_S:=\nu A+B$.

We subdivide the proof into three steps.

Step 1: Approximate local solutions
We start by constructing approximate solutions to our problem. To this end, we introduce smoothing operators J_k given by

$$J_k:=(1+k^{-1}A_S)^{-1},\quad k\in\mathbb{N}.$$

Since $-A_S$ is dissipative and generates a semigroup of contractions on $L^2_\sigma(\Omega)$, J_k is a bounded operator on $L^2_\sigma(\Omega)$ with $\|J_k\|_{L^2_\sigma(\Omega)}\le 1$ for all $k\in\mathbb{N}$ due

$$\|(1+k^{-1}A_S)^{-1}\|=\|k(k+A_S)^{-1}\|\le 1.$$

By Sobolev's embedding theorem we have for $u\in L^2_\sigma(\Omega)$

$$\|J_k u\|_\infty\le C(k)\|u\|_2. \tag{4.5}$$

Moreover, for every $1 < p < \infty$ there exist $k_0(p) \in \mathbb{N}$ and $C(p) > 0$, such that for $u \in L^p_\sigma(\Omega)$

$$\|J_k u\|_{L^p_\sigma(\Omega)} \leq C(p)\|u\|_{L^p_\sigma(\Omega)}, \quad k \geq k_0(p). \tag{4.6}$$

Indeed, this follows from the assumption that $-A_S$ generates an analytic semigroup on $L^p_\sigma(\Omega)$. Hence, in view of Remark 1.18, there exists $r > 0$, such that $r + A_S$ is a sectorial operator in $L^p_\sigma(\Omega)$ satisfying the corresponding resolvent estimates.

Now, we set

$$u_{0k} := J_k u_0$$

and

$$F_k u := -P(J_k u \cdot \nabla)u$$

and construct approximate solutions u_k to equation (4.2) by solving the integral equation

$$u_k(t) = e^{-tA_S}u_{0k} + \int_0^t e^{-(t-s)A_S}F_k u_k(s)\,\mathrm{d}s. \tag{4.7}$$

To this end, consider for $T > 0$ the Banach space $X := C([0,T], D(A_S^{\frac{1}{2}}))$, equipped with the norm

$$\|v\|_T := \sup_{0\leq t\leq T}(\|v(t)\|_2 + \|A_S^{\frac{1}{2}}v(t)\|_2)$$

and for $M > 0$ and $k \in \mathbb{N}$ the closed set

$$S(k, M, T) := \{v \in X, v(0) = u_{0k}, \|v\|_T \leq M\},$$

as well as the nonlinear operator Γ_k defined on $S(k, M, T)$ by

$$\Gamma_k v(t) := e^{-tA_S}u_{0k} + \int_0^t e^{-(t-s)A_S}F_k v(s)\,\mathrm{d}s.$$

Note that the estimate $\|F_k v(t)\|_2 \leq C(k)\|v(t)\|_2\|\nabla v(t)\|_2$ holds by (4.5). Since $D(A_S^{\frac{1}{2}}) = H_0^1(\Omega)^3 \cap L^2_\sigma(\Omega)$ by Remark 4.3, and therefore

$$\|\nabla v(t)\|_2 \leq \|v(t)\|_2 + \|A_S^{\frac{1}{2}}v(t)\|_2, \tag{4.8}$$

we obtain

$$\|F_k v(t)\|_2 \leq C(k)\big(\|v(t)\|_2^2 + \|v(t)\|_2\|A_S^{\frac{1}{2}}v(t)\|_2\big). \tag{4.9}$$

Since e^{-tA_S} is an analytic semigroup on $L^2_\sigma(\Omega)$, due to (4.4) and the Cauchy Schwartz inequality there is a $C > 0$, such that for all $u \in L^2_\sigma(\Omega)$

$$\|\nabla e^{-tA_S}u\|_2 \leq |\langle A_S e^{-tA_S}u, e^{-tA_S}u\rangle|^{\frac{1}{2}} \leq \|A_S e^{-tA_S}u\|_2^{\frac{1}{2}}\|e^{-tA_S}u\|_2^{\frac{1}{2}} \leq Ct^{-\frac{1}{2}}\|u\|_2$$

for $0 < t < T$. Thus, we may estimate $\Gamma_k v$ as follows:

$$\begin{aligned}
\|\Gamma_k v\|_T \leq\ & \|u_{0k}\|_2 + \|A_S^{\frac{1}{2}} u_{0k}\|_2 \\
& + \sup_{0\leq t\leq T} \{\int_0^t C(k)\|v(s)\|_2^2 \,\mathrm{d}s\} \\
& + \sup_{0\leq t\leq T} \{\int_0^t C(k)\|v(s)\|_2 \|A_S^{\frac{1}{2}} v(s)\|_2 \,\mathrm{d}s\} \\
& + \sup_{0\leq t\leq T} \{\int_0^t (t-s)^{-\frac{1}{2}} C(k)\|v(s)\|_2^2 \,\mathrm{d}s\} \\
& + \sup_{0\leq t\leq T} \{\int_0^t (t-s)^{-\frac{1}{2}} C(k)\|v(s)\|_2 \|A_S^{\frac{1}{2}} v(s)\|_2 \,\mathrm{d}s\} \\
\leq\ & \|u_{0k}\|_2 + \|A_S^{\frac{1}{2}} u_{0k}\|_2 + C_1(k) M^2 (T + T^{\frac{1}{2}}).
\end{aligned}$$

Furthermore, since for $u_1, u_2 \in S(k, M, T)$ and $t \in (0, T)$

$$\begin{aligned}
\|F_k u_1(t) - F_k u_2(t)\|_2 \leq\ & \|(J_k u_2(t) \cdot \nabla)(u_1 - u_2)(t)\|_2 \\
& + \|((J_k u_1 - J_k u_2)(t) \cdot \nabla) u_1(t)\|_2 \\
\leq\ & CM \|u_1 - u_2\|_T,
\end{aligned}$$

we obtain

$$\|\Gamma_k u_1 - \Gamma_k u_2\|_T \leq C_2(k) M (T + T^{\frac{1}{2}}) \|u_1 - u_2\|_T.$$

Now, fix M, such that $\|u_{0k}\|_2 + \|A_S^{\frac{1}{2}} u_{0k}\|_2 \leq \frac{M}{2}$, and then $T = T(k)$, such that $C_1(k) M^2 (T + T^{\frac{1}{2}}) \leq \frac{M}{2}$ and $C_2(k) M (T + T^{\frac{1}{2}}) < 1$. Then Γ_k is a strict contraction on $S(k, M, T)$ and by the Banach fixed point theorem, there exists a unique u_k in $S(k, M, T)$ satisfying (4.7) for $t \in (0, T)$.

Step 2: Approximate global solutions

In the following we prove a priori bounds for $u_k(T)$ and $A_S^{\frac{1}{2}} u_k(T)$ for all $T > 0$. To this end, recall that u_k is the solution of the equation

$$u_k'(t) + A_S u_k(t) = F_k u_k(t), \quad t \in (0, T). \tag{4.10}$$

Multiplying (4.10) with $u_k(t)$ and integrating by parts, yields

$$\frac{1}{2}\frac{\mathrm{d}}{\mathrm{d}t}\|u_k(t)\|_2^2 + \langle A_S u_k(t), u_k(t)\rangle = \langle F_k u_k(t), u_k(t)\rangle.$$

It follows from integration by parts that we have for $v, w \in D(A_S^{\frac{1}{2}})$

$$\begin{aligned}
\langle (w \cdot \nabla)v, v\rangle &= \int_\Omega \sum_{i=1}^{3}\sum_{j=1}^{3} w_i \partial_i v_j v_j \,\mathrm{d}x \\
&= -\int_\Omega \sum_{i=1}^{3}\sum_{j=1}^{3} w_i v_j \partial_i v_j \,\mathrm{d}x \\
&= -\langle (w \cdot \nabla v), v\rangle
\end{aligned}$$

and so $\langle F_k u_k(t), u_k(t)\rangle = 0$. Thus, condition (4.4) yields

$$\frac{1}{2}\frac{\mathrm{d}}{\mathrm{d}t}\|u_k(t)\|_2^2 + C\|\nabla u_k(t)\|_2^2 \le 0, \quad 0 < t < T, \tag{4.11}$$

for some $C > 0$. Note that (4.11) implies $\frac{\mathrm{d}}{\mathrm{d}t}\|u_k(t)\|_2^2 \le 0$ for $0 < t < T$. Therefore,

$$\|u_k(t)\|_2 \le \|u_{0k}\|_2, \quad 0 < t < T, \tag{4.12}$$

follows. Integrating (4.11) with respect to t yields

$$\|u_k(T)\|_2^2 + C\int_0^T \|\nabla u_k(s)\|_2^2 \,\mathrm{d}s \le \|u_0\|_2^2. \tag{4.13}$$

Next, forming the dual pairing of (4.10) with $A_S u_k(t)$,we obtain

$$\langle u_k'(t), A_S u_k(t)\rangle + \|A_S u_k(t)\|_2^2 = \langle F_k u_k(t), A_S u_k(t)\rangle.$$

Substituting $u_k'(t) = -A_S u_k(t) + F_k u_k(t)$, since $A_S u_k(t) = \nu A u_k(t) + B u_k(t)$ the equation

$$\begin{aligned}
\langle u_k'(t), A_S u_k(t)\rangle + \|A_S u_k(t)\|_2^2 &= \langle u_k'(t), \nu A u_k(t)\rangle - \langle A_S u_k(t), B u_k(t)\rangle \\
&\quad + \langle F_k u_k(t), B u_k(t)\rangle + \|A_S u_k(t)\|_2^2 \\
&= \langle F_k u_k(t), \nu A u_k(t)\rangle + \langle F_k u_k(t), B u_k(t)\rangle
\end{aligned}$$

follows. Now we may calculate

$$\begin{aligned}
\langle u_k'(t), \nu A u_k(t)\rangle &= \langle F_k u_k(t), \nu A u_k(t)\rangle + \langle A_S u_k(t), B u_k(t)\rangle - \|A_S u_k(t)\|_2^2 \\
&\le \langle F_k u_k(t), \nu A u_k(t)\rangle - \frac{1}{2}\|A_S u_k(t)\|_2^2 + \frac{1}{2}\|B u_k(t)\|_2^2 \\
&= \langle F_k u_k(t), A_S u_k(t)\rangle - \langle F_k u_k(t), B u_k(t)\rangle \\
&\quad - \frac{1}{2}\|A_S u_k(t)\|_2^2 + \frac{1}{2}\|B u_k(t)\|_2^2 \\
&\le \frac{1}{2}\|F_k u_k(t)\|_2^2 - \langle F_k u_k(t), B u_k(t)\rangle + \frac{1}{2}\|B u_k(t)\|_2^2 \\
&\le \|B u_k(t)\|_2^2 + \|F_k u_k(t)\|_2^2.
\end{aligned}$$

Thus, by condition (4.3) and inequality (4.5) this implies

$$\begin{aligned}\langle u_k'(t), \nu A u_k(t)\rangle &\le C(\|u_k(t)\|_2^2 + \|\nabla u_k(t)\|_2^2) + C(k)\|u_k(t)\|_2^2\|\nabla u_k(t)\|_2^2 \\ &\le C(\|u_0\|_2^2 + \|\nabla u_k(t)\|_2^2) + C(k)\|u_0\|_2^2\|\nabla u_k(t)\|_2^2,\end{aligned}$$

where the second inequality is due to the fact that $\|u_k(t)\|_2^2 \le \|u_{0k}\|_2^2 \le \|u_0\|_2^2$, which follows from (4.12) and the contractivity of J_k. Finally, since $\langle u_k'(t), \nu A u_k(t)\rangle = \frac{1}{2}\frac{\mathrm{d}}{\mathrm{d}t}\|\nabla u_k(t)\|_2^2$, integrating with respect to t yields together with (4.13)

$$\|\nabla u_k(T)\|_2^2 \le C(T\|u_0\|_2^2 + \|u_0\|_2^2) + C(k)\|u_0\|_2^4 + \|\nabla u_{0k}\|_2^2.$$

As $D(A_S^{\frac{1}{2}}) = H_0^1(\Omega)^3 \cap L^2_\sigma(\Omega)$ the estimate

$$\|A_S^{\frac{1}{2}} v\|_2 \le C(\|v\|_2 + \|\nabla v\|_2)$$

holds for every $v \in D(A_S^{\frac{1}{2}})$. Combining this with the estimate given in (4.13), we obtain for every $k \in \mathbb{N}$ and every $T > 0$ that $\|u_k\|_T < \infty$. Hence, u_k exists globally in time for all $k \in \mathbb{N}$.

Step 3: Convergence to a weak solution
In this final step we show that the approximate global solutions u_k constructed above converge in the weak sense to some weak solution u of system (4.2). We fix some time interval $[0, T]$. Since

$$\begin{aligned}\int_0^T \|A_S^{\frac{1}{2}} u_k(s)\|_2^2 \,\mathrm{d}s &\le 2\int_0^T \|u_k(s)\|_2^2 + \|\nabla u_k(s)\|_2^2 \,\mathrm{d}s \\ &\le 2T\|u_0\|_2^2 + 2\int_0^T \|\nabla u_k(s)\|_2^2 \,\mathrm{d}s,\end{aligned}$$

the above inequality (4.13) implies

$$u_k \in L^2((0,T), D(A_S^{\frac{1}{2}})) \cap L^\infty((0,T), L^2_\sigma(\Omega)) =: Y_1 \cap Y_2 =: Y, \quad k \in \mathbb{N}$$

and that (u_k) is even a bounded sequence in Y. Since Y_1 is reflexive, there exists a subsequence of (u_k) converging weakly in Y_1. Furthermore, by Alaoglu's theorem, (u_k) possesses a weak-star convergent subsequence in Y_2 and thus there exists a function $u \in Y$ with (u_k) converging weakly to u in Y_1 and (u_k) converging in the weak-star topology to u in Y_2.

Since u_k is a solution of the system

$$\left\{\begin{aligned} u_k'(t) + A_{SCE} u_k(t) + P(J_k u_k(t)\cdot\nabla)u_k(t) &= 0, \quad t > 0, \\ u_k(0) &= u_{0k}, \end{aligned}\right. \tag{4.14}$$

we may write $u_k(t) = u_k^{(1)}(t) + u_k^{(2)}(t)$ where

$$\begin{aligned} u_k^{(1)}(t) &:= e^{-tA_S} u_{0k}, \\ u_k^{(2)}(t) &:= \int_0^t e^{-(t-s)A_S} F_k u_k(s)\, \mathrm{d}s. \end{aligned}$$

Performing the same calculations, which led to (4.13), we now obtain for $k, l \in \mathbb{N}$

$$\|u_k^{(1)}(t) - u_l^{(1)}(t)\|_2^2 + \int_0^t \|A_S^{\frac{1}{2}}(u_k^{(1)}(s) - u_l^{(1)}(t))\|_2^2 \,\mathrm{d}s \le C\|u_{0k} - u_{0l}\|_2^2.$$

Since $u_{0k} \to u_0$ in $L^2_\sigma(\Omega)$ as $k \to \infty$, we see that $(u_k^{(1)})$ is a bounded sequence in Y. The same holds for $(u_k^{(2)})$, since (u_k) is bounded in Y as well.

Next, we set $r = \frac{5}{4}$. By Lemma 4.4, Hölder's and Sobolev's inequalities

$$\|F_k u_k(t)\|_r \le C\|J_k u_k(t)\|_{\frac{10}{3}} \|\nabla u_k(t)\|_2 \le C\|u_k(t)\|_2^{\frac{2}{5}} \|\nabla J_k u_k(t)\|_2^{\frac{3}{5}} \|\nabla u_k(t)\|_2.$$

Since we have by (4.8) that $\|\nabla J_k u_k(t)\|_2 \le \|A_S^{\frac{1}{2}} J_k u_k(t)\|_2 + \|J_k u_k(t)\|_2 \le C(\|A_S^{\frac{1}{2}} u_k(t)\|_2 + \|u_k(t)\|_2)$ and $\|\nabla u_k(t)\|_2 \le \|A_S^{\frac{1}{2}} u_k(t)\|_2 + \|u_k(t)\|_2$, we see that

$$\|F_k u_k(t)\|_r \le C\|u_0\|_2^{\frac{2}{5}} (\|A_S^{\frac{1}{2}} u_k(t)\|_2 + \|u_0\|_2)^{\frac{8}{5}}.$$

Furthermore, since (u_k) is bounded in Y_2, it follows from (4.13) that

$$\begin{aligned} \int_0^T \|F_k u_k(t)\|_r^r \,\mathrm{d}t &\le C\|u_0\|_2^{\frac{1}{2}} \int_0^T (\|A_S^{\frac{1}{2}} u_k(t)\|_2 + \|u_0\|_2)^2 \,\mathrm{d}t \\ &\le C(T+1)\|u_0\|_2^{\frac{5}{2}}. \end{aligned}$$

Hence, $(F_k u_k)$ is a bounded sequence in $L^r((0,T), L^r_\sigma(\Omega))$. By construction, $u_k^{(2)}$ is a solution of the Cauchy problem

$$\left\{ \begin{aligned} u_k'(t) + A_S u_k(t) &= F_k u_k(t), \quad t \ge 0, \\ u(0) &= 0. \end{aligned} \right.$$

By assumption, the operator A_S has maximal regularity in $L^r_\sigma(\Omega)$ on every finite time interval. Thus, $(u_k^{(2)})$ is a bounded sequence in $L^r((0,T), D(A_S)) \cap W^{1,r}((0,T), L^r_\sigma(\Omega))$. Due to (4.6) there is $k_0 \in \mathbb{N}$, such that for $k \ge k_0$ the operators J_k are uniformly bounded on $L^r_\sigma(\Omega)$. Thus, $(J_k u_k^{(2)})_{k \ge k_0}$ is a bounded sequence in $L^r((0,T), D(A_S)) \cap W^{1,r}((0,T), L^r_\sigma(\Omega))$ as well.

Since (u_k) is a bounded sequence in Y by (4.13) and since (u_k) and $(J_k u_k)$ are bounded in the space of maximal regularity, it follows from [Tem77, Theorem

III.2.1] and [Maz85, Lemma 1.4.6] that $(u_k^{(2)})$ and $(J_k u_k^{(2)})_{k\geq k_0}$ are relatively compact in $L^2(K\times(0,T))^3$ for any fixed compact set $K\subset\Omega$. It follows that there are subsequences $(u_{k_m}^{(2)})$ of $(u_k^{(2)})$ and $(J_{k_m}u_{k_m}^{(2)})$ of $(J_k u_k^{(2)})$, which converge in $L^2(\Omega\times(0,T))^3$. Therefore, there exists a subsequence (u_{k_m}) of (u_k), such that $u_{k_m}(s)\to u(s)$ and $J_{k_m}u_{k_m}(s)\to u(s)$ for a.a. $s\in(0,T)$ for some function $u\in Y$.

Finally, we need to verify that the function u constructed above is in fact a weak solution of our problem (4.2). To this end, note that by the weak convergence of (u_k) in Y_1, we have for all $\phi\in D(A^{\frac{1}{2}})$ and all $h\in C^1([0,T],\mathbb{R})$ with $h(T)=0$

$$\begin{aligned}
\lim_{k\to\infty}\int_0^T\langle u_k(t),\phi\rangle h'(t)\,\mathrm{d}t &= \int_0^T\langle u(t),\phi\rangle h'(t)\,\mathrm{d}t,\\
\lim_{k\to\infty}\int_0^T\langle\nabla u_k(t),\nabla\phi\rangle h(t)\,\mathrm{d}t &= \int_0^T\langle\nabla u(t),\nabla\phi\rangle h(t)\,\mathrm{d}t,\\
\lim_{k\to\infty}\int_0^T\langle\hat{B}u_k(t),\phi\rangle h(t)\,\mathrm{d}t &= \int_0^T\langle\hat{B}u(t),\phi\rangle h(t)\,\mathrm{d}t,\\
\lim_{k\to\infty}\langle u_{0k},\phi\rangle h(0), &= \langle u_0,\phi\rangle h(0)
\end{aligned}$$

since all these terms are linear. It remains to show that

$$\lim_{k\to\infty}\int_0^T\langle(J_k u_k(t)\cdot\nabla)u_k(t),\phi\rangle h(t)\,\mathrm{d}t=\int_0^T\langle(u(t)\cdot\nabla)u(t),\phi\rangle h(t)\,\mathrm{d}t.$$

Let $\chi_N:\Omega\to\{0,1\}$ be given by

$$\chi_N(x)=\begin{cases}1, & x\in\Omega\cap B(0,N),\\ 0, & \text{otherwise},\end{cases}$$

where $B(x,N)$ denotes the ball with center x and radius N, and consider

$$\int_0^T\langle(J_k u_k(t)\cdot\nabla)u_k(t),\phi\rangle h(t)\,\mathrm{d}t=I_1+I_2,$$

where

$$\begin{aligned}
I_1 &:= \int_0^T\langle(J_k u_k(t)\cdot\nabla)u_k(t),\chi_N\phi\rangle h(t)\,\mathrm{d}t,\\
I_2 &:= \int_0^T\langle(J_k u_k(t)\cdot\nabla)u_k(t),(1-\chi_N)\phi\rangle h(t)\,\mathrm{d}t.
\end{aligned}$$

Assume first that $\phi\in D(A_S^{\frac{1}{2}})\cap L^\infty$. Then

$$\lim_{k\to\infty}\int_0^T\langle(J_k u_k(t)\cdot\nabla)u_k(t),\chi_N\phi\rangle h(t)\,\mathrm{d}t=\int_0^T\langle(u(t)\cdot\nabla)u(t),\chi_N\phi\rangle h(t)\,\mathrm{d}t,$$

since $\chi_N\phi$ is bounded and has bounded support. If $\phi \notin L^\infty$, take $J_m\phi \in D(A_S^{\frac{1}{2}}) \cap L^\infty$ for some $m \in \mathbb{N}$ and then pass to the limit $m \to \infty$.

Since $n = 3$ we obtain by Hölder's and by Sobolev's inequality with $r = (\frac{1}{2} + \frac{1}{n})^{-1} = \frac{6}{5}$ and by the Gagliardo-Nirenberg inequality (1.11) that

$$\begin{aligned} &\Big|\int_0^T \langle (J_k u_k(t) \cdot \nabla) u_k(t), (1-\chi_N)\phi\rangle h(t)\, \mathrm{d}t\Big| \\ \leq\ & C\int_0^T \|(J_k u_k(t)\cdot\nabla)u_k(t)\|_r \|(1-\chi_N)\phi\|_{2^*}\, \mathrm{d}t \\ \leq\ & C\int_0^T \|J_k u_k(t)\|_n \|\nabla u_k(t)\|_2 \|(1-\chi_N)\phi\|_{2^*}\, \mathrm{d}t \\ \leq\ & C\int_0^T \|u_k(t)\|_2^{\frac{1}{2}} \|\nabla J_k u_k(t)\|_2^{\frac{1}{2}} \|\nabla u_k(t)\|_2 \|(1-\chi_N)\phi\|_{2^*}\, \mathrm{d}t, \end{aligned}$$

where $\frac{1}{2^*} = \frac{1}{2} - \frac{1}{n} = \frac{1}{6}$. Since $\phi \in D(A_S^{\frac{1}{2}})$ and $D(A_S^{\frac{1}{2}}) \hookrightarrow L^{2^*}(\Omega)$, we obtain

$$\lim_{N\to\infty}\Big|\int_0^T \langle (J_k u_k(t)\cdot\nabla)u_k(t), (1-\chi_N)\phi\rangle h(t)\, \mathrm{d}t\Big| = 0.$$

Since our choice of T was arbitrary in the construction above, for every $T > 0$ there exists a function $u \in L^\infty((0,T), L^2_\sigma(\Omega)) \cap L^2((0,T), D(A^{\frac{1}{2}}))$ satisfying assertion ii) of definition 4.1. For a given $T > 0$, the approximating functions u_k are uniquely determined. In particular, if v_k denotes the approximating solution on $(0, T_1)$ and w_k is the corresponding approximating solution on $(0, T_2)$, where $0 < T_1 < T_2$, then v_k and w_k coincide on $(0, T_1)$. Hence, for every $T > 0$ we find a limit solution u on $(0,T)$, which coincides with a limit solution $\tilde{u}$ on $(0, \tilde{T})$ for every $\tilde{T} < T$. This yields the existence of a weak solution $u : [0,\infty) \to L^2_\sigma(\Omega)$ of system (4.2).

□

Chapter 5

Asymptotic stability of the Ekman spiral in $\mathbb{R}^3_+$

In this chapter we will deal with the so-called Ekman spiral and prove results concerning its stability.

The Ekman spiral is an example of a mathematical model, which is originated in the geosciences. During an expedition to the nothern polar region the swedish oceanograph V.W. Ekman investigated the motion of icebergs swimming in the ocean. He observed that they do not move into the direction of the wind but receive a certain deflection. To find an explanation of this behaviour, he chose to model the ocean as the three-dimensional half-space $\mathbb{R}^3_+$ and considered the flow of the water to be described by the Navier-Stokes equations. Furthermore, he had to take into accout the effect of the Coriolis force, which is induced by the earth's rotation. Moreover, he assumed that there is no motion of the water at the ground of the ocean. Consequently, the mathematical model he had to deal with were the Navier-Stokes equations with additional Coriolis force term in $\mathbb{R}^3_+$ and homogeneous Dirichlet boundary conditions, which are given by

$$\left\{ \begin{array}{rcll} \partial_t u - \nu\Delta u + \omega(\mathbf{e}_3 \times u) & & & \\ + (u\cdot\nabla)u + \nabla p & = & 0, & t>0,\ x \in \mathbb{R}^3_+, \\ \operatorname{div} u & = & 0, & t>0,\ x \in \mathbb{R}^3_+, \\ u(t,x_1,x_2,0) & = & 0, & t>0,\ x_1,x_2 \in \mathbb{R}, \\ u(0,x) & = & u_0, & x \in \mathbb{R}^3_+, \end{array} \right. \tag{5.1}$$

where $u = (u^1, u^2, u^3)$ denotes the velocity field and p the pressure of an incompressible, viscous fluid. Here, $\mathbf{e}_3$ denotes the unit vector in x_3-direction, $\nu > 0$ the viscosity of the fluid, and the constant $\omega \in \mathbb{R}$ is called the Coriolis parameter, which is equal to twice the frequency of rotation around the x_3 axis.

Ekman was able to construct a time-independent solution to the above system (5.1), which can even be expressed explicitly as

$$\begin{cases} u_E(x_3) &= u_\infty(1 - e^{-x_3/\delta}\cos(x_3/\delta), e^{-x_3/\delta}\sin(x_3/\delta), 0)^T, \\ p_E(x_2) &= -\omega u_\infty x_2, \end{cases} \tag{5.2}$$

where δ is defined by $\delta := (\frac{2\nu}{\omega})^{1/2}$ and $u_\infty \geq 0$ is a constant. For details, see [Ekm05]. In his honour, this solution is called the *Ekman spiral.*

The Ekman spiral represents a mathematical description of rotating boundary layers in geophysical fluid dynamics (atmospheric and oceanic boundary layers) between a geostrophic flow and a solid boundary where the no slip boundary condition applies. In this setting the parameter δ denotes the thickness of the layer. It is a remarkable fact that the velocity field of this stationary solution depends on the space variable x_3 only. That means that the Ekman spiral corresponds to a model of a stratification of the according fluid. In the geostrophic flow region corresponding to large x_3, there is a uniform flow with velocity u_∞ in the x_1 direction. Associated with u_∞, a pressure gradient in the x_2-direction appears. The Ekman spiral in $\mathbb{R}^3_+$ matches this uniform velocity for large x_3 with the no slip boundary condition at $x_3 = 0$, i.e. we have $u_E(0) = 0$ and

$$u_E(x_3) \to (u_\infty, 0, 0) \quad \text{provided} \quad x_3 \to \infty.$$

We remark that $u_E, \nabla u_E \notin L^p(\mathbb{R}^3_+)$ for $1 \leq p < \infty$. On the other hand we have $u_E, \nabla u_E \in L^\infty(\mathbb{R}^3_+)$. Furthermore, the gradient of the Ekman spiral satisfies $|\nabla u_E| = O(e^{-x_3})$. The latter property will be essential for the approach we will use in the course of this chapter.

5.1 Results concerning stability of the Ekman spiral

Our focus in this chapter will be on the question of stability of the Ekman spiral for L^2-perturbations. In this context we consider solutions (v, q) of system (5.1), where the velocity field v forms a perturbation of the Ekman spiral by a function $v \in L^2_\sigma(\mathbb{R}^3_+)$. More precisely, we set

$$v := u - u_E, \quad \text{and} \quad q := p - p_E.$$

Since (u_E, p_E) is a stationary solution of (5.1), the pair (v, q) satisfies the equations

$$\left\{\begin{array}{rcll} \partial_t v - \nu\Delta v + \omega(\mathbf{e}_3 \times v) + (u_E \cdot \nabla)v & & & \\ + v_3\partial_3 u_E + (v\cdot\nabla)v + \nabla q & = & 0, & t>0,\ x\in\mathbb{R}^3_+, \\ \operatorname{div} v & = & 0, & t>0,\ x\in\mathbb{R}^3_+, \\ v(x_1,x_2,0) & = & 0, & t>0,\ x_1,x_2\in\mathbb{R}, \\ v(0,x) & = & v_0, & x\in\mathbb{R}^3_+, \end{array}\right. \tag{5.3}$$

where $v_0 = u_0 - u_E$.

In the course of this thesis, whenever we talk of *stability*, we mean stability in the Lyapunov sense. The basic technique that is used in our approach is to reduce a system of partial differential equations to an autonomous ordinary differential equation, which is posed in a Banach space valued environment. Such a system may be written as an initial value problem of the form

$$\left\{\begin{array}{rcll} y'(t) & = & f(y(t)), & t\in J, \\ y(0) & = & y_0. \end{array}\right. \tag{5.4}$$

Here $J = (0, T)$, $T > 0$, denotes a time interval and $f : X \to X$, where X is the phase space. In this context, stability means the following: consider an equilibrium u of (5.4) (i.e. u is a solution of (5.4), which is constant in t). In this case u is called *stable*, if for every $\varepsilon > 0$ there is a $\delta > 0$, such that for every solution v of the system

$$\left\{\begin{array}{rcll} x'(t) & = & f(t, y(t)), & t\in J, \\ x(0) & = & x_0, \end{array}\right.$$

where $\|x_0 - u\|_X < \delta$ we have $\|v(t) - u\|_X < \varepsilon$ for all $t \geq 0$. Moreover, if u is stable and we have

$$\lim_{t\to\infty} \|v(t) - u\|_X = 0,$$

we call u *asymptotically stable*. If u is not stable we say that it is *unstable*.

In this sense, the Ekman spiral is stable, if for every global solution $v : [0, \infty) \to L^2_\sigma(\mathbb{R}^3_+)$ of (5.3) we have that $\|v(t)\|_2$ is bounded. Furthermore, it is asymptotically stable, if

$$\lim_{t\to\infty} \|v(t)\|_2 = 0$$

holds.

A system of a moving fluid in a rotating framework is described by some non-dimensional characteristic number, the so-called *Reynolds number*. The Reynolds number corresponding to the Ekman spiral is defined by $\mathrm{Re} = u_\infty \delta \nu^{-1}$. In his work [Lil66] Lilly investigated the stability of the Ekman spiral using numerical methods. He found that instabilities appear, whenever the Reynols number exceeds a critical value of about $\mathrm{Re} \approx 55$. Following his observations, it seems reasonable to conjecture that there exists a critical Reynolds number Re_c with the property that, if $\mathrm{Re} < \mathrm{Re}_c$, then the perturbed nonlinear flow is stable, and that the flow is unstable provided $\mathrm{Re} > \mathrm{Re}_c$.

In this chapter we consider the problem of existence of global-in-time weak solutions to (5.3) and study in the following their nonlinear stability behaviour for initial data belonging to $L^2_\sigma(\mathbb{R}^3_+)$. We show in our first main result that there exists a weak solution to the above system (5.3) of nonlinear equations provided the Reynolds number is small enough. Secondly, assuming this condition, we prove that for every initial data $v_0 \in L^2_\sigma(\mathbb{R}^3_+)$ there exists at least one weak solution v to (5.3), such that

$$\lim_{t\to\infty} \|v(t)\|_2 = 0,$$

which shows in particular that the Ekman spiral is asymptotically stable with respect to L^2-perturbations. Moreover, it is even possible to estimate the decay rate. Indeed, we show that, if Re is small enough, $\alpha > 0$, and $v_0 \in L^2_\sigma(\mathbb{R}^3_+)$ satisfies

$$\|e^{-tA_{SCE}} v_0\|_2 = O(t^{-\alpha}),$$

then there exists at least one weak solution v to the nonlinear problem (5.1), such that $\|v(t)\|_2 \leq Ct^{-\tilde{\alpha}}$, where $\tilde{\alpha} = \min\{\alpha, \frac{1}{4}\}$. Here $e^{-tA_{SCE}}$ denotes the semigroup on $L^2_\sigma(\mathbb{R}^3_+)$ generated by the negative Stokes-Coriolis-Ekman operator defined in the following section.

The examination of stability of stationary solutions of systems like (5.1) in an L^2-framework seems to be a standard task. For example, Borchers and Miyakawa proved several results concerning stability of stationary solutions of the Navier-Stokes equations – without rotational effect – under L^2-perturbations. For details, see [BM92], [BM95]. As a basic ingredient of their proofs serves the property of the negative Stokes operator to generate a bounded analytic semigroup on L^2_σ over the respective domain. Taking into account the Coriolis force term, this property does no longer hold for the operator associated to the system (5.3). In particular, the standard gradient estimate for this semigroup, which is needed to treat the nonlinear term, fails. Furthermore, the strong stability of the semigroup cannot be obtained by the real characterization of analytic semigroups.

We remark that the critical value of the Reynolds number we find to ensure the nonlinear stability of the Ekman spiral is $\mathrm{Re}_c = \frac{1}{\sqrt{2}} \approx 1$, thus far away from the value Lilly found, and whose exceeding guarantees instability. Hence, there remains a large gap in the range of the Reynolds number, where we cannot predict the stability behaviour of the Ekman spiral.

The question of stability of the Ekman spiral has been subject to the work of several authors. The first systematic examination was done in [Lil66] where the author obtained a critical value of the Reynolds number where an exceeding implies instability.

Analytical results on linear instability of v for large Reynolds number are more difficult to obtain. It was shown in [DG03] that in the case of flows between infinite layers, there exists a sequence of approximate solutions to (5.3), which is nonlinearly unstable for sufficiently large Reynolds numbers in the sense of [DG03]. Moreover, they show that, if the linearized version

$$\left\{\begin{array}{rcll} \partial_t v - \nu\Delta v + \omega(\mathbf{e}_3 \times v) & & & \\ + (u_E \cdot \nabla)v + v_3\partial_3 u_E + \nabla q & = & 0, & t>0,\ x \in \mathbb{R}^3_+, \\ \operatorname{div} v & = & 0, & t>0,\ x \in \mathbb{R}^3_+, \\ v(x_1,x_2,0) & = & 0, & t>0,\ x_1,x_2 \in \mathbb{R}, \\ v(0,x) & = & v_0, & x \in \mathbb{R}^3_+, \end{array}\right.$$

of system (5.3) admits instability, then the solution of the nonlinear equations also does.

Recently, it was shown in [GIM⁺07] that the nonlinear equation (5.3) admits a unique, *local* mild solution for all non decaying initial data belonging to a certain Besov space.

In the article [DDG99] the authors treat so-called Ekman-Hartmann boundary layers. These boundary layers appear in magnetohydrodynamic flows in a rotating framework. The limit case of a vanishing magnetic field corresponds to the Navier-Stokes equations with Coriolis force term.

They consider the magnetohydrodynamic equations

$$\left\{\begin{array}{rcll} \partial_t u - \frac{E}{\varepsilon}\Delta u + \frac{\mathbf{e}_3 \times u}{\varepsilon} & & & \\ + (u\cdot\nabla)u + \frac{\nabla p}{\varepsilon} & = & \frac{\Lambda\theta}{\varepsilon}\operatorname{curl} b \times b, & t>0,\ x \in \mathbb{R}^3_+, \\ \partial_t b + u\cdot\nabla b & = & b\cdot\nabla u + \frac{\Delta b}{\theta}, & t>0,\ x \in \mathbb{R}^3_+, \\ \operatorname{div} b & = & 0, & t>0,\ x \in \mathbb{R}^3_+, \\ \operatorname{div} u & = & 0, & t>0,\ x \in \mathbb{R}^3_+, \\ u(0,x) & = & u_0, & x \in \mathbb{R}^3_+, \\ b(0,x) & = & b_0, & x \in \mathbb{R}^3_+, \end{array}\right. \tag{5.5}$$

where u is the velocity of a viscous fluid, p its pressure, and b the magnetic field. According to the system the Ekman number E, the Rossby number ε, the Elsasser number Λ, and the magnetic Reynolds number θ are defined. The velocity field u is assumed to converge for $x_3 \to \infty$ to a constant limit flow u_∞, which is independent of x_3. Then the corresponding Reynolds number of the boundary layer is defined as $\mathrm{Re}_{BL} = |u_\infty|\frac{\varepsilon}{\sqrt{E}}$.

The geophysical situation they want to model is the earth's core. By physical reasons, in this setting the Rossby number ε and the Ekman number E are very small ($\varepsilon \approx 10^{-7}$, $E \approx 10^{-15}$). Furthermore, they assume that in general $E \simeq \varepsilon^2$ holds. These considerations bring the authors to investigate convergence of solutions of (5.5) when the parameter ε tends to 0 whereas the Elsasser number Λ is constant. In particular, they assume the existence of global-in-time weak solutions u^ε of system (5.5) corresponding to the respective value of ε. Then, for well-prepared initial data (for details see [DDG99]) they prove that in the limiting process $\varepsilon \to 0$ these solutions u^ε converge to a solution u^e of the two-dimensional damped Euler equation

$$\begin{aligned} \partial_t u^e + (u^e \cdot \nabla) u^e + \beta u^e + \nabla p^e &= 0, \\ \operatorname{div} u^e &= 0, \end{aligned}$$

where β is a constant depending on the Elsasser number Λ.

Moreover, they construct time independent solutions (u_s, b_s) of (5.5). Concerning the stability of these solutions they prove the following result:

Proposition 5.1. *The Ekman-Hartmann layers are stable provided*

$$\mathrm{Re}_{BL} \leq \mathrm{Re}_s(\Lambda).$$

More precisely, under this condition, if (u, b) is another solution of (5.5), then

$$\sup_{t \geq 0} \int_{\mathbb{R}^3_+} \left(|u(t) - u_s|^2 + \frac{\Lambda\theta}{\varepsilon} |b(t) - b_s|^2 \right) \leq \int_{\mathbb{R}^3_+} \left(|u_0 - u_s|^2 + \frac{\Lambda\theta}{\varepsilon} |b_0 - b_s|^2 \right).$$

The value $\mathrm{Re}_s(\Lambda)$ is explicitly given in the article.

Breaking down the result to the pure Ekman case (i.e. the magnetic field b vanishes) this means that for a solution v of (5.3) they obtain

$$\|v(t)\|_2 \leq \|v_0\|_2$$

for all $t \geq 0$. This result corresponds to the notion of stability in the Lyapunov sense, which was introduced above.

In this thesis, we prove a stronger result. In particular, we give a detailed proof of the existence of global-in-time weak solutions of system (5.3) and obtain explicit decay rates for these solutions.

For more information on the Navier-Stokes equations with rotational effect, we refer to [Gre80] or [BMN01].

5.2 Main result

Applying the Helmholtz projection P, we may rewrite equation (5.3) as an evolution equation in $L^p_\sigma(\mathbb{R}^3_+)$ of the form

$$\begin{cases} v'(t) + A_{SCE}v(t) + P(v(t)\cdot\nabla)v(t) &= 0, \quad t>0, \\ v(0) &= v_0, \end{cases} \tag{5.6}$$

where the Stokes-Coriolis-Ekman operator A_{SCE} in $L^p_\sigma(\mathbb{R}^3_+)$ is defined by

$$\begin{cases} A_{SCE}v &:= P(-\nu\Delta v + \omega(\mathbf{e}_3 \times v) + [(u_E\cdot\nabla)v + v_3\partial_3 u_E]) \\ &= (A + A_C + A_{E,1} + A_{E,2})v \\ D(A_{SCE}) &:= W^{2,p}(\mathbb{R}^3_+)^3 \cap W^{1,p}_0(\mathbb{R}^3_+)^3 \cap L^p_\sigma(\mathbb{R}^3_+). \end{cases} \tag{5.7}$$

We will now state our main result concerning weak solutions of (5.3) and nonlinear stability of the Ekman spiral.

Theorem 5.2. *Assume that $u_\infty\delta\nu^{-1} < \frac{1}{\sqrt{2}}$. Then the following assertions hold.*

a) There exists a weak solution to (5.3).

b) For every $v_0 \in L^2_\sigma(\mathbb{R}^3_+)$ there exists at least one weak solution v of (5.3), such that

$$\lim_{t\to\infty} \|v(t)\|_2 = 0.$$

c) Assume that for $v_0 \in L^2_\sigma(\mathbb{R}^3_+)$ and some $\alpha > 0$

$$\|e^{-tA_{SCE}}v_0\|_2 = O(t^{-\alpha}).$$

Then there exists at least one weak solution v of (5.3), such that

$$\|v(t)\|_2 = \begin{cases} O(t^{-\alpha}), & \alpha \le \frac{1}{4}, \\ O(t^{-\frac{1}{4}}), & \alpha > \frac{1}{4}. \end{cases}$$

The proof of Theorem 5.2 requires several preparations that we will carry out in the next sections.

5.3 The Stokes-Coriolis-Ekman semigroup on $L^2_\sigma(\mathbb{R}^3_+)$

In this section we show that $-A_{SCE}$ is the generator of an analytic semigroup in $L^p_\sigma(\mathbb{R}^3_+)$. Moreover, we prove that $-A_{SCE}$ generates a contraction semigroup on $L^2_\sigma(\mathbb{R}^3_+)$ provided the Reynolds number $\mathrm{Re} = u_\infty\delta\nu^{-1}$ is small enough. In this case the semigroup is even strongly stable.

Theorem 5.3. *Let $1 < p < \infty$. Then there is a $\mu > 0$, such that the operator $\mu + A_{SCE}$ has maximal regularity in $L^p_\sigma(\mathbb{R}^3_+)$. In particular, $-A_{SCE}$ is the generator of an analytic semigroup.*

Proof. In view of the characterization Theorem 1.31 it would be sufficient to show that the operator $\mu + A_{SCE}$ is $\mathcal{R}$-sectorial for some suitable $\mu > 0$. Therefore we have to show that $(A_C + A_{E,1} + A_{E,2})$ can be absorbed in terms of the shifted Stokes operator $(\mu + A)$. That means that there exist $\alpha > 0$ and $\mu > 0$, such that

$$\|(A_C + A_{E,1} + A_{E,2})v\|_p \leq \alpha\|(\mu + A)v\|_p$$

for every $v \in D(A)$. As before we immediately obtain that $D(A_C + A_{E,1} + A_{E,2}) \supset D(A)$. From the resolvent estimate given in Proposition 2.1 it follows that there is a $M > 0$, such that for every $v \in D(A)$ and all $\lambda \in \rho(A)$

$$|\lambda|\|v\|_p + |\lambda|^{\frac{1}{2}}\|\nabla v\|_p + \|\nabla^2 v\|_p \leq M\|(\lambda + A)v\|_p \tag{5.8}$$

holds. Furthermore, the results of [DHP01] (see (2.4)) imply that for every angle $0 < \theta < \pi$ there exists

$$a := \mathcal{R}\{A(\lambda + A)^{-1} : \lambda \in \Sigma_\theta\}.$$

Note that

$$\|(A_C + A_{E,1} + A_{E,2})v\|_p \leq \|A_C v\|_p + \|A_{E,1}v\|_p + \|A_{E,2}v\|_p.$$

In the following we will treat the three terms on the right hand side separately. Since

$$\|A_C v\|_p \leq \|P\||\omega|\|v\|_p$$

we obtain by (5.8) that $\|A_C v\|_p \leq \alpha\|(\mu + A)v\|_p$, whenever $\mu > \frac{M\|P\||\omega|}{\alpha}$, where M denotes the constant appearing in (5.8). Looking at the second term we find that

$$\|A_{E,1}v\|_p \leq \|P\||u_\infty|\|\nabla v\|_p.$$

Using (5.8) we see that $\|A_{E,1}v\|_p \leq \alpha\|(\mu + A)v\|_p$ for $\mu > \left(\frac{M\|P\||u_\infty|}{\alpha}\right)^2$. Since

$$\|A_{E,2}v\|_p \leq \|P\||u_\infty|\frac{1}{\delta}\|v\|_p,$$

we get as for A_C that $\|A_{E,2}v\|_p \leq \alpha\|(\mu + A)v\|_p$ for all $\mu > \frac{M\|P\||u_\infty|}{\alpha\delta}$. Put $\alpha = \frac{1}{a}$ and set

$$m := \max\left\{M\|P\||\omega|a, (M\|P\||u_\infty|a)^2, M\|P\||u_\infty|a\frac{1}{\delta}\right\}.$$

Then we obtain by Proposition 1.30 that $\mu + A_{SCE}$ is an $\mathcal{R}$-sectorial operator, whenever $\mu > 3m$. In particular, there exists a constant $C > 0$, such that for $1 < p, q < \infty$ and a solution u of the abstract Cauchy problem associated to A_{SCE} the inequality

$$\|u'\|_{L^q((0,\infty),L^p_\sigma(\mathbb{R}^3_+))} + \|(\mu + A_{SCE})u\|_{L^q((0,\infty),L^p_\sigma(\mathbb{R}^3_+))} \leq C\|f\|_{L^q((0,\infty),L^p_\sigma(\mathbb{R}^3_+))}$$

holds. □

Theorem 5.4. *The negative Stokes-Coriolis-Ekman operator $-A_{SCE}$ generates a C_0-semigroup of contractions on $L^2_\sigma(\mathbb{R}^3_+)$, provided*

$$\frac{u_\infty \delta}{\nu} \leq \frac{1}{\sqrt{2}}. \tag{5.9}$$

Note that by Theorem 5.3 we already know that $-A_{SCE}$ is the generator of an analytic C_0-semigroup $e^{-tA_{SCE}}$ on $L^p_\sigma(\mathbb{R}^3_+)$ for all $1 < p < \infty$. The uniform boundedness of $e^{-tA_{SCE}}$ in $L^2_\sigma(\mathbb{R}^3_+)$ will be essential in the following.

Lemma 5.5. *Let $p \in \mathbb{N}$ and $\alpha > 0$. Then there exists a constant $C(p) > 0$, such that*

$$\|e^{-(\cdot)/\alpha} v(\cdot)\|_{L^p(\mathbb{R}_+)} \leq C(p)\alpha \|v'\|_{L^p(\mathbb{R}_+)}$$

for $\alpha > 0$ and all $v \in W^{1,p}_0(\mathbb{R}_+)$. In the case $p = 2$ we have $C(2) = 1/2$.

Proof. Note first that it suffices to show the assertion for $v \in C^\infty_c(\mathbb{R}_+)$. In this case the result then follows from the representation

$$e^{-s/\alpha} v(s) = e^{-s/\alpha} \int_0^s v'(t)\, \mathrm{d}t, \quad s > 0.$$

In fact, due to the Jensen inequality we have

$$\begin{aligned} \|e^{-(\cdot)/\alpha} v(\cdot)\|^p_{L^p(\mathbb{R}_+)} &= \int_0^\infty e^{-ps/\alpha} \left| \int_0^s v'(t)\, \mathrm{d}t \right|^p \mathrm{d}s \\ &\leq \int_0^\infty e^{-ps/\alpha} s^{p-1} \int_0^s |v(t)|^p\, \mathrm{d}t\, \mathrm{d}s \\ &\leq \|v'\|^p_{L^p(\mathbb{R}_+)} \int_0^\infty e^{-ps/\alpha} s^{p-1}\, \mathrm{d}s. \end{aligned}$$

By integration by parts, for $p \in \mathbb{N}$ we simply obtain

$$\int_0^\infty e^{-ps/\alpha} s^{p-1}\, \mathrm{d}s = \alpha^p \frac{(p-1)!}{p^p},$$

which proves the assertion. □

Proof of Theorem 5.4. For $v_0 \in L^2_\sigma(\mathbb{R}^3_+)$ set $v(t) := e^{-tA_{SCE}}v_0$. Then v satisfies

$$\begin{cases} v'(t) + A_{SCE}v(t) &= 0, \quad t > 0, \\ v(0) &= v_0. \end{cases}$$

Multiplying the above equation with $v(t)$ and taking into account the skew symmetry of the second and third term of A_{SCE} we obtain

$$\frac{1}{2}\frac{\mathrm{d}}{\mathrm{d}t}\int_{\mathbb{R}^3_+} |v(t)|^2 \,\mathrm{d}x + \nu \int_{\mathbb{R}^3_+} |\nabla v(t)|^2 \,\mathrm{d}x + \int_{\mathbb{R}^3_+} v(t)\cdot(v_3(t)\cdot\partial_3 u_E)\,\mathrm{d}x = 0, \quad t > 0.$$

Since

$$\int_{\mathbb{R}^3_+} v(t)\cdot(v_3(t)\cdot\partial_3 u_E)\,\mathrm{d}x \le \sum_{j=1}^{2} \|e^{(\cdot)/2\delta}(\partial_3 u_E)_j v_3(t)\|_2 \|e^{-(\cdot)/2\delta} v_j(t)\|_2$$

and since

$$\partial_3 u_E(x_3) = \frac{u_\infty}{\delta} e^{-x_3/\delta} \begin{pmatrix} \cos(x_3/\delta) + \sin(x_3/\delta) \\ \cos(x_3/\delta) - \sin(x_3/\delta) \\ 0 \end{pmatrix},$$

we see that

$$\|e^{(\cdot)/2\delta}(\partial_3 u_E)_j v_3(t)\|_2 \le \sqrt{2}\frac{u_\infty}{\delta}\|e^{-(\cdot)/2\delta} v_3(t)\|_2. \tag{5.10}$$

The above Lemma 5.5 implies

$$\|e^{-(\cdot)/2\delta} v_j(t)\|_2 \le \delta\|\partial_3 v_j(t)\|_2, \quad j = 1, 2, 3.$$

Combining these estimates, we finally have

$$\int_{\mathbb{R}^3_+} v(t)\cdot(v_3(t)\cdot\partial_3 u_E)\,\mathrm{d}x \le \sqrt{2}u_\infty\delta\|\nabla v(t)\|_2^2.$$

Thus,

$$\frac{\mathrm{d}}{\mathrm{d}t}\|v(t)\|_2^2 \le 0 \tag{5.11}$$

for all $t > 0$, provided

$$u_\infty\delta \le \frac{\nu}{\sqrt{2}}.$$

Therefore

$$\|e^{-tA_{SCE}}(t)v_0\|_2 = \|v(t)\|_2 \le \|v_0\|_2, \quad t > 0,$$

and the assertion is proved.

□

As a main ingredient for the stability estimates of weak solutions of system (5.6) we will use the following result concerning the strong stability of the semigroup generated by $-A_{SCE}$ in $L^2_\sigma(\mathbb{R}^3_+)$.

Theorem 5.6. *Suppose the parameters u_∞, ν, and δ fulfill condition (5.9) with strict inequality. Then for every $v \in L^2_\sigma(\mathbb{R}^3_+)$ we have*

$$\lim_{t\to\infty} \|e^{-tA_{SCE}}v\|_2 = 0, \tag{5.12}$$

i.e. the semigroup generated by $-A_{SCE}$ in $L^2_\sigma(\mathbb{R}^3_+)$ is strongly stable.

Proof. By the fact that now the Reynolds number is strictly less than $1/\sqrt{2}$, the same calculation as performed in the proof of Theorem 5.4 implies that

$$\frac{1}{2}\frac{\mathrm{d}}{\mathrm{d}t}\|v(t)\|_2^2 + C\|\nabla v(t)\|_2^2 \le 0,$$

for some $C > 0$. Integrating with respect to t yields

$$\|v(T)\|_2^2 + C\int_0^T \|\nabla v(t)\|_2^2 \,\mathrm{d}t \le \|v_0\|_2^2$$

for all $T \in (0,\infty)$. In particular, this estimate implies

$$\int_0^\infty \|\nabla e^{-tA_{SCE}}v\|_2^2 \,\mathrm{d}t \le C\|v\|_2^2 \tag{5.13}$$

for every $v \in L^2_\sigma(\mathbb{R}^3_+)$. In what follows we will concentrate on the derivative w.r.t. x_1, i.e., on the estimate

$$\int_0^\infty \|\partial_1 e^{-tA_{SCE}}v\|_2^2 \,\mathrm{d}t \le C\|v\|_2^2, \quad v \in L^2_\sigma(\mathbb{R}^3_+). \tag{5.14}$$

In the next step assume that $v \in D(A^{3/2})$. As $e^{-tA_{SCE}}$ is a semigroup of contractions and since ∂_1 commutes with all appearing operators it follows that

$$\begin{aligned}\left|\frac{\mathrm{d}}{\mathrm{d}t}\|e^{-tA_{SCE}}\partial_1 v\|_2^2\right| &= 2|\langle e^{-tA_{SCE}}A_{SCE}\partial_1 v, e^{-tA_{SCE}}\partial_1 v\rangle| \\ &\le 2\|A_{SCE}\partial_1 v\|_2\|\partial_1 v\|_2 \\ &\le 2\|v\|^2_{D(A^{3/2})}.\end{aligned}$$

Thus, since $t \mapsto e^{-tA_{SCE}}$ is a strongly continuous mapping, we have $\|e^{-tA_{SCE}}A^{1/2}v\|_2^2 \in C^1[0,\infty) \cap W^{1,\infty}(0,\infty)$. Combining this with (5.14), we obtain

$$\lim_{t\to\infty} \|\partial_1 e^{-tA_{SCE}}v\|_2 = 0 \tag{5.15}$$

for every $v \in D(A^{3/2})$. Moreover, since $D(A^{3/2})$ is dense in $D(A^{1/2})$, for every $v \in D(A^{1/2})$ there is a sequence $(v_n)_{n\in\mathbb{N}} \subset D(A^{3/2})$, such that $v_n \to v$ in $D(A^{1/2})$. It follows that

$$\|\partial_1 e^{-tA_{SCE}}v\|_2 \le \|\partial_1 e^{-tA_{SCE}}(v - v_n)\|_2 + \|\partial_1 e^{-tA_{SCE}}v_n\|_2.$$

Thus, for given $\varepsilon > 0$ we may choose n large enough, such that $\|e^{-tA_{SCE}}\partial_1(v - v_n)\|_2 \leq C\|v - v_n\|_{H^1} < \frac{\varepsilon}{2}$, and then $t > 0$, such that $\|\partial_1 e^{-tA_{SCE}} v_n\|_2 < \frac{\varepsilon}{2}$. Hence, relation (5.15) is true even for $v \in D(A^{1/2})$.

Next, observe that the operator

$$Q_2 := (1 + A)^{-1} P(1 - \Delta)$$

obviously is a projection from $D(-\Delta) = H^2(\mathbb{R}^3_+) \cap H^1_0(\mathbb{R}^3_+)$ onto $D(A)$. By a duality argument it is not difficult to see that Q_2 has a bounded extension from $L^2(\mathbb{R}^3_+)$ onto $L^2_\sigma(\mathbb{R}^3_+)$, which we denote by Q_0 (and which is not the Helmholtz projection). Complex interpolation and the fact that A admits a bounded $\mathcal{H}^\infty$-calculus on $L^2_\sigma(\mathbb{R}^3_+)$ then yields a projection

$$Q_1 : D((-\Delta)^{1/2}) \to D(A^{1/2}),$$

which is surjective and which is also extended to $L^2_\sigma(\mathbb{R}^3_+)$ through Q_0. Employing the projections Q_1 and Q_0 relation (5.15) can be written as

$$\lim_{t\to\infty} \|e^{-tA_{SCE}}\partial_1 Q_1 v\|_2 = \lim_{t\to\infty} \|e^{-tA_{SCE}} Q_0 \partial_1 v\|_2 = 0, \quad v \in D((-\Delta)^{1/2}).$$

In view of $D((-\Delta)^{1/2}) = H^1_0(\mathbb{R}^3_+)$ the above relation is in particular true for all $v \in H^1(\mathbb{R}, H^1_0(\mathbb{R}^2_+))$, where the exterior space H^1 represents the space for the x_1-variable. By the fact that $H^1(\mathbb{R}, H^1_0(\mathbb{R}^2_+))$ is dense in $H^1(\mathbb{R}, L^2(\mathbb{R}^2_+))$ a similar $\varepsilon/2$-argument as above shows that

$$\lim_{t\to\infty} \|e^{-tA_{SCE}} Q_0 \partial_1 v\|_2 = 0$$

for all $v \in H^1(\mathbb{R}, L^2(\mathbb{R}^2_+))$. It is well-known that

$$\partial_1 : H^1(\mathbb{R}, L^2(\mathbb{R}^2_+)) \to L^2(\mathbb{R}^3_+)$$

is skew-selfadjoint and injective in $L^2(\mathbb{R}, L^2(\mathbb{R}^2_+)) = L^2(\mathbb{R}^3_+)$. From this we see that also $\partial_1' = -\partial_1$ is injective, and therefore ∂_1 has dense range. By a similar $\varepsilon/2$-argument as above we conclude that

$$\lim_{t\to\infty} \|e^{-tA_{SCE}} Q_0 v\|_2 = 0, \quad v \in L^2_\sigma(\mathbb{R}^3_+),$$

which yields the assertion. □

5.4 Existence of weak solutions

In this section we prove the existence of a weak solution to problem (5.6) (hence, to problem (5.3), too) in the case where the Reynolds number $\mathrm{Re} = u_\infty \delta \nu^{-1}$ is small enough. We therefore assume throughout this section that

$$\mathrm{Re} = \frac{u_\infty \delta}{\nu} < \frac{1}{\sqrt{2}}.$$

Basically, all we have to do is to check, whether the operator A_{SCE} satisfies the conditions of Theorem 4.2. Then the claim follows directly from this result.

Lemma 5.7. *Consider the Stokes-Coriolis-Ekman operator*

$$\begin{aligned} A_{SCE}v &= -\nu P\Delta v + \omega P(\mathbf{e}_3 \times v) + (P(u_E \cdot \nabla)v + Pv_3\partial_3 u_E) \\ &= (A + A_C + A_{E,1} + A_{E,2})v. \end{aligned}$$

Setting $B := A_C + A_{E,1} + A_{E,2}$, all the conditions stated in Theorem 4.2 are satisfied.

Proof. 1. The operator $A_C + A_{E,1} + A_{E,2}$ is the sum of bounded operators and a first order differential operator. Thus, it is closed, if we choose its domain as $D(A_C + A_{E,1} + A_{E,2}) = W_0^{1,p}(\mathbb{R}^3_+)^3 \cap L^p_\sigma(\mathbb{R}^3_+)$, which is obviously a superset of $D(A)$. Furthermore, using the graph norm in $H^1_0(\mathbb{R}^3_+)^3 \cap L^2_\sigma(\mathbb{R}^3_+)$, estimate (4.3) holds.

2. Taking into Theorem 5.3, we immediately obtain the assertion.

3. By Theorem 5.4 the given choice of the Reynolds number assures that for $v \in L^2_\sigma(\mathbb{R}^3_+)$

$$\|\nabla v\|_2^2 \leq \mathrm{Re}\,\langle A_{SCE}v, v\rangle.$$

□

Having Lemma 5.7 at hand, we obtain immediately assertion a) of Theorem 5.2 by Theorem 4.2.

5.5 Proof of the stability estimates

In the foregoing section we showed the existence of weak solutions of system (5.6). In this section we will prove the claims concerning the decay of these solutions given in Theorem 5.2.

Proof of assertions b) and c) of Theorem 5.2.

We recall that we constructed approximate solutions v_k, which converge to some weak solution v of problem (5.6). Each of these approximate solutions satisfies

$$v_k(t) = e^{-tA_{SCE}}v_{0k} - \int_0^t e^{-(t-s)A_{SCE}}P(J_k v_k(s) \cdot \nabla)v_k(s)\,\mathrm{d}s \tag{5.16}$$

for all $t > 0$. In the following we will prove stability estimates of these approximate solutions, which are uniformly in k and transfer to the solution v of the original problem.

Note that $\nabla e^{-tA_{SCE}^*}$ is a bounded operator on $L^2_\sigma(\mathbb{R}^3_+)$ for every $t \in (0,\infty)$. Hence, the nonlinear term may be estimated as

$$\begin{aligned}
& \|e^{-(t-s)A_{SCE}}P(J_k v_k(t)\cdot\nabla)v_k(t)\|_2 \\
= & \sup_{\phi\in L^2_\sigma(\mathbb{R}^3_+),\|\phi\|_2=1} |\langle e^{-(t-s)A_{SCE}}P(J_k v_k(t)\cdot\nabla)v_k(t),\phi\rangle| \\
= & \sup_{\phi\in L^2_\sigma(\mathbb{R}^3_+),\|\phi\|_2=1} |\langle J_k v_k(t)\otimes v_k(t), \nabla e^{-(t-s)A_{SCE}^*}\phi\rangle| \\
\le & \|v_k(t)\otimes v_k(t)\|_2 \|\nabla e^{-(t-s)A_{SCE}^*}\| \\
\le & \|v_k(t)\|_4^2 \|\nabla e^{-(t-s)A_{SCE}^*}\|.
\end{aligned}$$

Since the dimension of the underlying space is $n = 3$, we infer from the Gagliardo-Nirenberg inequality (1.11) that

$$\|v_k(t)\|_4^2 \le C\|v_k(t)\|_2^{\frac{1}{2}}\|\nabla v_k(t)\|_2^{\frac{3}{2}}.$$

By (5.16) these estimates give

$$\|v_k(t)\|_2 \le \|e^{-tA_{SCE}}v_{0k}\|_2 + C\int_0^t \|\nabla e^{-(t-s)A_{SCE}^*}\|\cdot\|v_k(s)\|_2^{\frac{1}{2}}\|\nabla v_k(s)\|_2^{\frac{3}{2}}\,\mathrm{d}s. \tag{5.17}$$

We set

$$\begin{aligned}
F_1(t) &:= \int_0^t \|\nabla e^{-(t-s)A_{SCE}^*}\|\cdot\|v_k(s)\|_2^2\,\mathrm{d}s, \\
F_2(t) &:= \int_0^t \|\nabla e^{-(t-s)A_{SCE}^*}\|\cdot\|\nabla v_k(s)\|_2^2\,\mathrm{d}s.
\end{aligned}$$

Then, due to Hölder's inequality,

$$\begin{aligned}
& \int_0^t \|\nabla e^{-(t-s)A_{SCE}^*}\|\cdot\|v_k(s)\|_2^{\frac{1}{2}}\|\nabla v_k(s)\|_2^{\frac{3}{2}}\,\mathrm{d}s \\
= & \int_0^t \|\nabla e^{-(t-s)A_{SCE}^*}\|^{\frac{1}{4}}\|v_k(s)\|_2^{\frac{1}{2}}\cdot\|\nabla e^{-(t-s)A_{SCE}^*}\|^{\frac{3}{4}}\|\nabla v_k(s)\|_2^{\frac{3}{2}}\,\mathrm{d}s \\
\le & \left(\int_0^t (\|\nabla e^{-(t-s)A_{SCE}^*}\|^{\frac{1}{4}}\|v_k(s)\|_2^{\frac{1}{2}})^4\,\mathrm{d}s\right)^{\frac{1}{4}} \\
& \cdot\left(\int_0^t (\|\nabla e^{-(t-s)A_{SCE}^*}\|^{\frac{3}{4}}\|\nabla v_k(s)\|_2^{\frac{3}{2}})^{\frac{4}{3}}\,\mathrm{d}s\right)^{\frac{3}{4}} \\
= & (F_1(t))^{\frac{1}{4}}\cdot(F_2(t))^{\frac{3}{4}}.
\end{aligned}$$

holds. By (4.11), $\frac{\mathrm{d}}{\mathrm{d}t}\|v_k(t)\|_2 \le 0$ for all $t > 0$. Therefore we have

$$\frac{\mathrm{d}}{\mathrm{d}t}\|v_k(t)\|_2 + \frac{1}{t}\|v_k(t)\|_2 \le \frac{1}{t}\|v_k(t)\|_2. \tag{5.18}$$

Applying (5.17) to the right-hand side of (5.18) and multiplying this inequality by t yields

$$\frac{\mathrm{d}}{\mathrm{d}t}(t\|v_k(t)\|_2) \leq \|e^{-tA_{SCE}}v_{0k}\|_2 + C(F_1(t))^{\frac{1}{4}}(F_2(t))^{\frac{3}{4}}, \quad t > 0.$$

Denoting the second term on the right hand side by $F(t)$, we obtain after integrating in t and dividing by t

$$\|v_k(t)\|_2 \leq t^{-1}\int_0^t \|e^{-\tau A_{SCE}}v_{0k}\|_2 \,\mathrm{d}\tau + t^{-1}\int_0^t F(\tau)\,\mathrm{d}\tau. \tag{5.19}$$

Furthermore, using Hölder's inequality, we calculate

$$\begin{aligned} t^{-1}\int_0^t F(\tau)\,\mathrm{d}\tau &= Ct^{-1}\int_0^t (F_1(\tau))^{\frac{1}{4}}(F_2(\tau))^{\frac{3}{4}}\,\mathrm{d}\tau \\ &\leq Ct^{-1}\Big(\int_0^t F_1(\tau)\,\mathrm{d}\tau\Big)^{\frac{1}{4}}\Big(\int_0^t F_2(\tau)\,\mathrm{d}\tau\Big)^{\frac{3}{4}} \\ &= C\Big(t^{-1}\int_0^t F_1(\tau)\,\mathrm{d}\tau\Big)^{\frac{1}{4}}\Big(t^{-1}\int_0^t F_2(\tau)\,\mathrm{d}\tau\Big)^{\frac{3}{4}}. \end{aligned} \tag{5.20}$$

Recall that we have due to inequality (5.13)

$$\int_0^t \|\nabla e^{-sA_{SCE}^*}\|^2\,\mathrm{d}s \leq C$$

for all $t \in [0,\infty)$. Hence, we obtain by the Cauchy-Schwartz inequality

$$\int_0^t \|\nabla e^{-sA_{SCE}^*}\|\,\mathrm{d}s \leq t^{\frac{1}{2}}\Big(\int_0^t \|\nabla e^{-sA_{SCE}^*}\|^2\,\mathrm{d}s\Big)^{\frac{1}{2}} \leq Ct^{\frac{1}{2}}.$$

Since $t \mapsto \|v_k(t)\|_2^2 \in L^\infty(0,\infty)$ by (4.12), it follows that

$$\begin{aligned} t^{-1}\int_0^t F_1(\tau)\,\mathrm{d}\tau &= t^{-1}\int_0^t\int_0^\tau \|\nabla e^{-(\tau-s)A_{SCE}^*}\|\cdot\|v_k(s)\|_2^2\,\mathrm{d}s\,\mathrm{d}\tau \\ &\leq Ct^{-1}\int_0^t\int_s^t (\tau-s)^{\frac{1}{2}}\,\mathrm{d}\tau\|v_k(s)\|_2^2\,\mathrm{d}s \\ &\leq Ct^{-1}\int_0^t (t-s)^{\frac{1}{2}}\,\mathrm{d}s \\ &\leq Ct^{\frac{1}{2}}. \end{aligned}$$

On the other hand, by (4.13) we have that $t \mapsto \|\nabla v_k(t)\|_2^2 \in L^1(0,\infty)$. Thus, we obtain

$$\begin{aligned}
t^{-1}\int_0^t F_2(\tau)\,\mathrm{d}\tau &= t^{-1}\int_0^t\int_0^\tau \|\nabla e^{-(\tau-s)A_{SCE}^*}\|\cdot\|\nabla v_k(s)\|_2^2\,\mathrm{d}s\,\mathrm{d}\tau \\
&\leq t^{-1}\int_0^t\int_s^t (\tau-s)^{\frac{1}{2}}\,\mathrm{d}\tau\|\nabla v_k(s)\|_2^2\,\mathrm{d}s \\
&\leq Ct^{-1}\int_0^t (t-s)^{\frac{1}{2}}\|\nabla v_k(s)\|_2^2\,\mathrm{d}s \\
&\leq Ct^{-1}\cdot t^{\frac{1}{2}}\cdot\int_0^t \|\nabla v_k(s)\|_2^2\,\mathrm{d}s \\
&\leq Ct^{-\frac{1}{2}}.
\end{aligned}$$

Applying these estimates of F_1 and F_2 in (5.20) yields

$$\|v_k(t)\|_2 \leq t^{-1}\int_0^t \|e^{-\tau A_{SCE}}v_{0k}\|_2\,\mathrm{d}\tau + Ct^{-\frac{1}{4}}. \tag{5.21}$$

By Theorem 5.6 we observe

$$\lim_{t\to\infty}\|e^{-tA_{SCE}}v_{0k}\|_2 = 0.$$

Then estimate (5.21) implies $\lim_{t\to\infty}\|v_k(t)\|_2 = 0$. If we assume in addition that $\|e^{-tA_{SCE}}v_0\|_2 = O(t^{-\alpha})$ for some $\alpha > 0$, then we have that even $t^{-1}\int_0^t \|e^{-\tau A_{SCE}}v_{0k}\|_2\,\mathrm{d}\tau = O(t^{-\alpha})$ holds. Since these estimates are uniformly in k, we have proved assertions b) and c) of Theorem 5.2.

□

Chapter 6

Stability of the Ekman spiral in infinite layers

This chapter will be devoted to the stability behaviour of the Ekman spiral in infinite layers instead of the halfspace $\mathbb{R}^3_+$. By an infinite layer we mean a domain, which is given by

$$\Omega_b := \{x \in \mathbb{R}^3 : 0 < x_3 < b\}. \tag{6.1}$$

for some positive number b.

The choice to consider infinite layers as the underlying space where the motion of the respective fluid takes place is justified by physical reasons. We recall that the primal model, for which the Ekman spiral was developed, is the motion of water in an ocean. Another field in geophysical research where the concept of the Ekman spiral is frequently used is the mathematical modeling of atmospheric flows. The latter models are an essential ingredient for modern methods of weather forecast and the prediction of atmospheric turbulences, which are a major threat for the safety of aircrafts. The archetypes for these models (ocean and atmosphere) have natural bounds in the vertical direction. In the case of the ocean we have the ground as the lower bound and the water surface as the upper bound. In case of the atmosphere we have to keep in mind that they are developed for the investigation of weather incidents. All these phenomena occur in the troposphere only, which is the lowest part of the earth's atmosphere. Thus, we have as natural bounds of this region the earth's surface as the lower bound and the tropopause, which is the border between the troposphere and the higher located stratosphere, as the upper bound.

The above examples show that infinite layers are a much more reasonable approach than the halfspace to model such areas, in which geophysical phenomena take place. In fact, in simulations of geophysical flows often an three-dimensional infinite layer is taken as a basis. For example, we refer to the recent work [TCY05] where this method is used.

In the following we will construct an explicit solution to the Navier-Stokes equations with rotational effect in an infinite layer. For this purpose, we make use of the already known explicit representation of the Ekman spiral in $\mathbb{R}^3_+$. We consider a domain Ω_b as defined in (6.1). The boundary of Ω_b consists of two components and will further be described by

$$\partial\Omega_b = \Gamma_b^1 \cap \Gamma_b^2,$$

where

$$\Gamma_b^1 := \{x \in \mathbb{R}^3 : x_3 = 0\}$$

and

$$\Gamma_b^2 := \{x \in \mathbb{R}^3 : x_3 = b\}.$$

We would like to model the situation of a fluid in a rotating infinite layer, where on the lower part of the boundary Γ_b^1 a homogeneous Dirichlet boundary condition applies, whereas on the upper part Γ_b^2 the fluid moves with a constant velocity u_b in the x_1 direction. This behaviour seems to be reasonable due to the geophysical situations, whose modeling we keep in mind. For example, in the case of an ocean we may assume a moving water surface, which is driven by wind and where the no-slip condition applies. On the other hand, if we look at atmospheric flows, a constant velocity at the upper bound corresponds to the jet stream. Hence, the velocity and the pressure of the fluid have to fulfill the system of equations

$$\left\{\begin{array}{rcll} \partial_t u - \nu\Delta u + \omega(\mathbf{e}_3 \times u) & & & \\ + (u\cdot\nabla)u + \nabla p & = & 0, & t>0,\ x\in\Omega_b, \\ \operatorname{div} u & = & 0, & t>0,\ x\in\Omega_b, \\ u(t,x_1,x_2,0) & = & 0, & t>0,\ x_1,x_2\in\mathbb{R}, \\ u(t,x_1,x_2,b) & = & \mathbf{e}_1\cdot u_b, & t>0,\ x_1,x_2\in\mathbb{R}, \\ u(0) & = & u_0, & x\in\Omega_b. \end{array}\right. \tag{6.2}$$

The Ekman spiral

$$u_E = \begin{pmatrix} u_\infty(1-e^{-\frac{x_3}{\delta}}\cos\left(\frac{x_3}{\delta}\right)) \\ u_\infty(e^{-\frac{x_3}{\delta}}\sin\left(\frac{x_3}{\delta}\right)) \\ 0 \end{pmatrix}$$

as defined in chapter 5 is an explicitly given time independent solution of the Navier-Stokes equations with Coriolis term in the halfspace. Assume now that δ is given by

$$\delta = \frac{b}{k\pi} \tag{6.3}$$

for some $k \in \mathbb{Z}$. We define

$$\tilde{u}_b := \begin{cases} u_b \left(1 - e^{-\frac{b}{\delta}}\right)^{-1}, & \text{if } k \text{ is even,} \\ u_b \left(1 + e^{-\frac{b}{\delta}}\right)^{-1}, & \text{if } k \text{ is odd.} \end{cases} \tag{6.4}$$

Let

$$u_b^E(x_3) := \begin{pmatrix} \tilde{u}_b(1 - e^{-\frac{x_3}{\delta}} \cos\left(\frac{x_3}{\delta}\right)) \\ \tilde{u}_b(e^{-\frac{x_3}{\delta}} \sin\left(\frac{x_3}{\delta}\right)) \\ 0 \end{pmatrix} \tag{6.5}$$

and

$$p_b^E(x_2) := -\omega \tilde{u}_b x_2. \tag{6.6}$$

Since $u_b^E(b) = (u_b, 0, 0)^T$, we obtain that (u_b^E, p_b^E) is a stationary solution of (6.2).

Now we are interested in stability of the solution (u_b^E, p_b^E). That means that we consider perturbations of (u_b^E, p_b^E) by functions u and p solving system (6.2). To this end, set

$$v := u - u_b^E, \quad \text{and} \quad q := p - p_b^E.$$

Since (u_b^E, p_b^E) is a stationary solution of (6.2), the pair (v, q) satisfies the equations

$$\begin{cases} \partial_t v - \nu \Delta v + \omega(\mathbf{e}_3 \times v) + (u_b^E \cdot \nabla)v & \\ \qquad + v_3 \partial_3 u_b^E + (v \cdot \nabla)v + \nabla q = 0, & t > 0,\ x \in \Omega_b, \\ \operatorname{div} v = 0, & t > 0,\ x \in \Omega_b, \\ v(t, x_1, x_2, 0) = v(t, x_1, x_2, b) = 0, & t > 0,\ x_1, x_2 \in \mathbb{R}, \\ v(0) = v_0, & x \in \Omega_b, \end{cases} \tag{6.7}$$

where $v_0 = u_0 - u_b^E$. Note that in (6.7) we have homogeneous Dirichlet boundary conditions on Γ_b in contrast to (6.2) where homogeneous boundary values were assumed only on Γ_b^1.

In this chapter we will focus our investigations on the solutions of equations (6.7). In a first step we will show that the results we obtained in chapter 5 for the halfspace case transfer to the case of infinite layers. That means that in the case of small Reynolds numbers there exists a weak solution $v : [0, \infty) \to L^2_\sigma(\Omega_b)$, such that $\lim_{t \to \infty} \|v(t)\|_2 = 0$. We are able to show that this solution actually exhibits an exponential decay. Furthermore, the particular properties of the evolution operator associated to (6.7) in $L^p_\sigma(\Omega_b)$ (see section 6.1) allow us to give a proof for the existence of strong solutions $v : [0, T) \to L^3_\sigma(\Omega_b)$ of equations (6.7), where $T \in (0, \infty]$ depends on v_0.

We remark that these results are substantially stronger than the ones we could obtain in the halfspace case. In particular, in the halfspace all results we presented on stability are vaild in the Hilbert space setting only. In fact, even the boundedness of the semigroup generated by the negative Stokes-Coriolis-Ekman operator in $L^p_\sigma(\mathbb{R}^3_+)$ is an open problem for $p \neq 2$. Moreover, in the halfspace it is not known whether so-called global L^p-L^q estimates hold for this semigroup. Thus, the basic ingredient for the construction of *strong* solutions to system (5.3) is missing.

Our approach fundamentally relies on the technique to regard the Stokes-Coriolis-Ekman operator as a perturbation of the corresponding Stokes operator. Hence, the question what results we are able to obtain is closely related to the properties of the Stokes operator in the respective domain. The Stokes operator in $L^p_\sigma(\Omega_b)$ for $1 < p < \infty$ is very well investigated. In [AS03a] and [Abe05a] the unique solvability of the Stokes resolvent equations as well as resolvent estimates of the according solution were proved for resolvent parameters λ, that are contained in a sector Σ_θ for $0 < \theta < \pi$. Since the infinite layer is bounded in one coordinate, we have that $0 \in \rho(A)$. In [AS03b] Abe and Shibata transferred the resolvent estimates to the case $\lambda = 0$. In [AW05] Abels and Wiegner improved this result and obtained the unique solvability and the corresponding resolvent estimates for all resolvent parameters $\lambda \in \mathbb{C} \setminus (-\infty, -\pi^2/b^2)$, where $b > 0$ denotes the layers thickness. Moreover, their result includes global L^p-L^q estimates of the solution of the Stokes resolvent equation. The most recent result on the Stokes operator in infinite layers is given in [Abe05b]. There it is shown that the Stokes operator A has maximal regularity and admits a bounded $\mathcal{H}^\infty$-calculus in $L^p_\sigma(\Omega_b)$ for $1 < p < \infty$.

By spectral properties of the Stokes operator in infinite layers it follows that the semigroup generated by $-A$ is exponentially stable. More precisely, there are constants $C, \mu > 0$, such that $\|e^{-tA}\| \leq Ce^{-t\mu}$ for all $t \geq 0$. In Section 6.3, we will transfer the property of exponential decay to global-in-time weak L^2-solutions of (6.7). Furthermore, since we have $0 \in \rho(A)$, it is possible to show that even the semigroup generated by the linear operator associated to (6.7) admits global L^p-L^q estimates. These estimates are the key for the construction of strong solutions of system (6.7) in Section 6.4.

6.1 The Stokes-Coriolis-Ekman operator in infinite layers

Applying the Helmholtz projection P from, we may rewrite equation (6.7) as an evolution equation in $L^p_\sigma(\Omega_b)$ of the form

$$\left\{ \begin{array}{rcll} v'(t) + A_{SCE}v(t) + P(v(t) \cdot \nabla)v(t) & = & 0, & t > 0, \\ v(0) & = & v_0, & \end{array} \right. \tag{6.8}$$

where the Stokes-Coriolis-Ekman operator A_{SCE} in $L^p_\sigma(\Omega_b)$ is defined by

$$\begin{cases} A_{SCE}v & := & P(-\nu\Delta v + \omega(\mathbf{e}_3 \times v) + [(u_b^E \cdot \nabla)v + v_3\partial_3 u_b^E]) \\ & = & (A + A_C + A_E)v, \\ D(A_{SCE}) & := & W^{2,p}(\Omega_b)^3 \cap W_0^{1,p}(\Omega_b)^3 \cap L^p_\sigma(\Omega_b). \end{cases} \tag{6.9}$$

It is shown in [Abe05b] that for $1 < p < \infty$ the Stokes operator A in $L^p_\sigma(\Omega_b)$ admits maximal L^q-regularity. In order to prove that this holds true also for the Stokes-Coriolis-Ekman operator, in view of Proposition 1.30 and Proposition 1.31, we have to show that $A_{SCE} - A$ is relatively bounded with respect to A. By the same arguments as in the proof of Lemma 5.7 we may observe that there exists a constant $\mu > 0$, such that

$$\mathcal{R}\{\lambda(\lambda + \mu + A_{SCE})^{-1} : \lambda \in \Sigma_\theta\} < \infty.$$

Thus, Proposition 1.30 implies the following result.

Proposition 6.1. *Let $1 < p, q < \infty$, $f \in L^q((0,T), L^p_\sigma(\Omega_b))$ and A_{SCE} in $L^p_\sigma(\Omega_b)$ be given as in (6.9). Then there exists $\mu > 0$, such that $\mu + A_{SCE}$ admits maximal L^q-regularity on $L^p_\sigma(\Omega_b)$. In particular, there exists a constant $C > 0$, such that*

$$\|u'\|_{L^q((0,T),L^p_\sigma(\Omega_b))} + \|(\mu + A_{SCE})u\|_{L^q((0,T),L^p_\sigma(\Omega_b))} \le C\|f\|_{L^q((0,T),L^p_\sigma(\Omega_b))}$$

holds. In particular, $-A_{SCE}$ generates an analytic semigroup on $L^p_\sigma(\Omega_b)$.

An essential ingredient in the course of this chapter will be that $-A_{SCE}$ generates a semigroup of contractions on $L^2_\sigma(\Omega_b)$ in the case of small Reynolds numbers.

Theorem 6.2. *Let the parameters $\tilde{u}_b$, ν, b, and δ fulfill the condition*

$$\frac{\tilde{u}_b}{\nu}\left(\delta - be^{-\frac{b}{\delta}} - \delta e^{-\frac{b}{\delta}}\right) \le \frac{1}{\sqrt{2}}. \tag{6.10}$$

Then the operator $-A_{SCE}$ is the generator of a semigroup of contractions on $L^2_\sigma(\Omega_b)$.

Proof. Let $v_0 \in L^2_\sigma(\Omega_b)$. Then $v(t) = e^{-tA_{SCE}}v_0$ satisfies

$$\begin{cases} v'(t) + A_{SCE}v(t) & = & 0, \quad t > 0, \\ v(0) & = & v_0. \end{cases}$$

Multiplying the first line with $v(t)$ and integrating over Ω_b, we obtain

$$\frac{1}{2}\frac{\mathrm{d}}{\mathrm{d}t}\|v(t)\|_2^2 + \nu\|\nabla v(t)\|_2^2 + \langle v_3(t) \cdot \partial_3 u_b^E, v(t)\rangle = 0,$$

where the second and the third term of A_{SCE} disappear due to their skew symmetry. Using

$$\partial_3 u_b^E = \frac{\tilde{u}_b}{\delta} e^{-\frac{x_3}{\delta}} \begin{pmatrix} \cos(x_3/\delta) + \sin(x_3/\delta) \\ \cos(x_3/\delta) - \sin(x_3/\delta) \\ 0 \end{pmatrix}$$

and applying the fundamental theorem of calculus, we may estimate

$$|\langle v_3(t) \cdot \partial_3 u_b^E, v(t)\rangle| \leq \frac{\sqrt{2}\tilde{u}_b}{\delta} \int_{\Omega_b} \left| \int_0^{x_3} \partial_3 v(t, x_1, x_2, \xi) \, \mathrm{d}\xi \right|^2 e^{-\frac{x_3}{\delta}} \, \mathrm{d}x.$$

Furthermore, the Cauchy-Schwarz inequality yields

$$\left| \int_0^{x_3} \partial_3 v(t, x_1, x_2, \xi) \, \mathrm{d}\xi \right|^2 \leq x_3 \|\partial_3 v(t, x_1, x_2, \cdot)\|_{L^2(0,b)}^2.$$

Finally, since $\|\partial_3 v(t, x_1, x_2, \cdot)\|_{L^2(0,b)}^2$ does not depend on x_3 we have

$$\int_{\Omega_b} x_3 \|\partial_3 v(t, x_1, x_2, \cdot)\|_{L^2(0,b)}^2 e^{-\frac{x_3}{\delta}} \, \mathrm{d}x \leq \|\nabla v(t)\|_2^2 \int_0^b x_3 e^{-\frac{x_3}{\delta}} \, \mathrm{d}x_3.$$

Hence, by the above estimates, the inequality

$$|\langle v_3(t) \cdot \partial_3 u_b^E, v(t)\rangle| \leq \sqrt{2}\tilde{u}_b \left(\delta - be^{-\frac{b}{\delta}} - \delta e^{-\frac{b}{\delta}}\right) \|\nabla v(t)\|_2^2 \tag{6.11}$$

holds. Thus, in case (6.10) is satisfied we have

$$\frac{\mathrm{d}}{\mathrm{d}t} \|v(t)\|_2^2 \leq 0$$

for every $t > 0$. Therefore, we have

$$\|e^{-tA_{SCE}} v_0\|_2 = \|v(t)\|_2 \leq \|v_0\|_2, \quad t > 0,$$

which proves the assertion. □

A minor restriction of assumption (6.10) entails that the semigroup generated by $-A_{SCE}$ is exponentially stable. As a preparation for the proof of this fact, we state a result concerning the Dirichlet Laplacian on layers Ω_b.

Proposition 6.3. *Let Ω_b be defined as in (6.1). Consider the Dirichlet Laplacian in $L^p(\Omega_b)$, i.e. $D(\Delta) = W^{2,p}(\Omega_b) \cap W_0^{1,p}(\Omega_b)$. Then*

$$\sigma(\Delta) \subset (-\infty, -\frac{\pi^2}{b^2}].$$

For a proof we refer to [AW05].

In the following let $\lambda_0 := -\frac{\pi^2}{b^2}$ denote the least upper bound of the spectrum of the Dirichlet Laplacian in $L^2(\Omega_b)$.

Lemma 6.4. *Let $\tilde{u}_b$, ν, b, and δ fulfill the condition*

$$\frac{\tilde{u}_b}{\nu}\left(\delta - be^{-\frac{b}{\delta}} - \delta e^{-\frac{b}{\delta}}\right) < \frac{1}{\sqrt{2}}. \tag{6.12}$$

Then there is $\mu > 0$, such that

$$\|e^{-tA_{SCE}}\| \le e^{-t\mu}, \quad t > 0.$$

Proof. By the results of Section 6.1, $-A_{SCE}$ is the generator of an analytic semigroup. Now, for $\mu > 0$ consider the operator $\mu - A_{SCE}$, which is the generator of an analytic semigroup, too. Due to the skew symmetry of the second and the third term of A_{SCE}, we have for all $v \in D(A_{SCE})$

$$\begin{aligned} \operatorname{Re}\langle(\mu - A_{SCE})v, v\rangle &= \mu\|v\|_2^2 - \nu\|\nabla v\|_2^2 + \operatorname{Re}\langle v_3 \cdot \partial_3 u_b^E, v\rangle \\ &\le \mu\|v\|_2^2 - \nu\|\nabla v\|_2^2 + |\langle v_3 \cdot \partial_3 u_b^E, v\rangle|. \end{aligned}$$

The first eigenvalue of the Dirichlet Laplacian may be represented by the Rayleigh-Ritz formula by

$$\lambda_0 = \sup_{v\in D(\Delta), \|v\|_2=1} \langle \Delta v, v\rangle = \sup_{v \in H_0^1(\Omega_b), v\neq 0} -\frac{\|\nabla v\|_2^2}{\|v\|_2^2}. \tag{6.13}$$

Since $\lambda_0 = -\frac{\pi^2}{b^2}$ this implies

$$\|v\|_2^2 \le \frac{b^2}{\pi^2}\|\nabla v\|_2^2, \quad v \in H_0^1(\Omega_b). \tag{6.14}$$

Moreover, by the same calculations as in the proof of Theorem 6.2 we obtain

$$|\langle v_3 \cdot \partial_3 u_b^E, v\rangle| < \sqrt{2}\tilde{u}_b\left(\delta - be^{-\frac{b}{\delta}} - \delta e^{-\frac{b}{\delta}}\right)\|\nabla v\|_2^2. \tag{6.15}$$

Note that the inequality is strict due to assumption (6.12). Since (6.12) is a strict inequality, we may choose $\mu > 0$, such that

$$\mu\frac{b^2}{\pi^2\nu} + \frac{\tilde{u}_b}{\nu}\left(\delta - be^{-\frac{b}{\delta}} - \delta e^{-\frac{b}{\delta}}\right) < \frac{1}{\sqrt{2}}.$$

Then, combining inequalities (6.14) and (6.15), we obtain that $\operatorname{Re}\langle(\mu - A_{SCE})v, v\rangle < 0$, which means that $\mu - A_{SCE}$ is dissipative in $L^2_\sigma(\Omega_b)$. By Theorem 1.12, this implies that $e^{t(\mu - A_{SCE})}$ is a semigroup of contractions. This yields

$$\|e^{t(\mu - A_{SCE})}\| = e^{t\mu}\|e^{-tA_{SCE}}\| \le 1,$$

which proves the assertion for $e^{-tA_{SCE}}$. □

6.2 Weak L^2-solutions in infinite layers

This section will be devoted to the existence of weak solutions to equations (6.8). For this purpose, we consider the Stokes-Coriolis-Ekman operator A_{SCE} in $L^p_\sigma(\Omega_b)$ as a perturbation of the Stokes operator A. We observe that by the results of the foregoing section we already know that $-A_{SCE}$ is the generator of a contractive semigroup on $L^2_\sigma(\Omega_b)$. Hence, in view of Proposition 1.22 its square root $A_{SCE}^{\frac{1}{2}}$ is well defined. Thus, we may ask for the existence of weak solutions to (6.8) in the sense of Definition 4.1.

It will be sufficient again to verify the conditions of Theorem 4.2, which immediately implies the existence of a weak solution in the sense of Definition 4.1.

Lemma 6.5. *Consider the operator*

$$\begin{aligned} A_{SCE}v &= -\nu P\Delta v + \omega P(\mathbf{e}_3 \times v + (u_E \cdot \nabla)v + v_3\partial_3 u_E) \\ &= (A + A_C + A_{E,1} + A_{E,2})v \end{aligned}$$

and let the parameters $\tilde{u}_b$, ν, b, and δ fulfill condition (6.10). Setting $B := A_C + A_{E,1} + A_{E,2}$, all the conditions stated in Theorem 4.2 are satisfied.

Proof. 1. Analogously to the halfspace case we have that $A_C + A_{E,1} + A_{E,2}$ is the sum of bounded operators and a first order differential operator. Hence $D(A_C + A_{E,1} + A_{E,2})$ may be chosen as $W^{1,p}_0(\Omega_b)^3 \cap L^p_\sigma(\Omega_b)$. So we have $D(A_C + A_{E,1} + A_{E,2}) \supset D(A)$. Using the graph norm in $H^1_0(\Omega_b)^3 \cap L^2_\sigma(\Omega_b)$, estimate (4.3) holds.

2. By Proposition 6.1 we obtain that for every $1 < p < \infty$ there is a $\mu > 0$, such that $\mu + A_{SCE}$ has maximal regularity on $L^p_\sigma(\Omega_b)$. Hence, the second condition of Theorem 4.2 is fulfilled.

3. Since condition (6.10) is satisfied, inequality (6.11) yields that for all $u \in D(A_{SCE})$

$$0 \leq \operatorname{Re}\langle A_{SCE}u, u\rangle$$

holds. □

Corollary 6.6. *Let $v_0 \in L^2_\sigma(\Omega_b)$ and condition (6.10) be fulfilled. Then there exists a weak solution v of (6.8) (hence, of (6.7)).*

6.3 Decay of weak solutions

Having proved the existence of a weak L^2-solutions of system (6.8), we now focus on the long-time behaviour of this solution.

Theorem 6.7. *We consider problem (6.7).*

1. *Let* $\lambda < 0$. *Furthermore, let the parameters* ν, δ, $\tilde{u}_b$, *and* λ *fulfill the condition*
$$\nu + \frac{b^2}{\pi^2}\lambda - \sqrt{2}\tilde{u}_b\left(\delta - be^{-\frac{b}{\delta}} - \delta e^{-\frac{b}{\delta}}\right) > 0. \tag{6.16}$$
Then for the weak solution v *obtained from Corollary 6.6*
$$\|v(t)\|_2 \le e^{\lambda t}\|v_0\|_2, \quad t \ge 0,$$
holds.

2. *Let* $\lambda \in \mathbb{R}$. *Furthermore, let the parameters* ν, δ, $\tilde{u}_b$, *and* λ *fulfill the condition*
$$\lambda + \frac{\pi^2}{b^2}\nu - \sqrt{2}\frac{\tilde{u}_b}{\delta} > 0. \tag{6.17}$$
Then for the weak solution v
$$\|v(t)\|_2 \le e^{\lambda t}\|v_0\|_2, \quad t \ge 0,$$
holds.

Proof. Recall that the weak solution v of (6.7) is the limit of a sequence of approximate solutions v_k, each being a solution of the smoothed system

$$\begin{cases} v_k'(t) + A_{SCE}v_k(t) + P(J_k v_k(t)\cdot\nabla)v_k(t) &= 0, \quad t>0, \\ v_k(0) &= v_{0k}, \end{cases} \tag{6.18}$$

where $J_k = (1 + k^{-1}A_{SCE})^{-1}$. In the following, we will show the stability estimates for all v_k. Since these estimates are uniformly in k, they transfer to the weak solution v.

We start by forming the dual pairing of the first line of (6.18) with $e^{-2\lambda t}v_k$ to obtain

$$\int_{\Omega_b} v_k'(t)e^{-2\lambda t}v_k(t)\,\mathrm{d}x + \nu\|e^{-\lambda t}\nabla v_k(t)\|_2^2 + e^{-2\lambda t}\langle v_{k3}(t)\partial_3 u_b^E, v_k(t)\rangle = 0$$

for all $t > 0$. Here we used that for all $u \in D(A_{SCE})$ we have

$$\langle (u\cdot\nabla)u, u\rangle = \langle (u_b^E\cdot\nabla)u, u\rangle = \langle \omega(\mathbf{e}_3\times u), u\rangle = 0$$

due to the skew symmetry of these terms. Now, the equation

$$\int_{\Omega_b} v_k'(t)e^{-2\lambda t}v_k(t)\,\mathrm{d}x = \frac{1}{2}\frac{\mathrm{d}}{\mathrm{d}t}\|e^{-\lambda t}v_k(t)\|_2^2 + \lambda e^{-2\lambda t}\int_{\Omega_b} v_k^2(t)\,\mathrm{d}x, \quad t>0,$$

yields

$$\begin{aligned} 0 &= \frac{1}{2}\frac{\mathrm{d}}{\mathrm{d}t}\|e^{-\lambda t}v_k(t)\|_2^2 \\ &\quad + \lambda e^{-2\lambda t}\|v_k(t)\|_2^2 + e^{-2\lambda t}\nu\|\nabla v_k(t)\|_2^2 + e^{-2\lambda t}\langle v_{k3}(t)\partial_3 u_b^E, v_k(t)\rangle \\ &= \frac{1}{2}\frac{\mathrm{d}}{\mathrm{d}t}\|e^{-\lambda t}v_k(t)\|_2^2 \\ &\quad + e^{-2\lambda t}\left(\lambda\|v_k(t)\|_2^2 + \nu\|\nabla v_k(t)\|_2^2 + \langle v_{k3}(t)\partial_3 u_b^E, v_k(t)\rangle\right). \end{aligned}$$

Let $\lambda < 0$. We recall that

$$|\langle v_{k3}(t)\partial_3 u_b^E, v_k(t)\rangle| \le \sqrt{2}\tilde{u}_b\left(\delta - be^{-\frac{b}{\delta}} - \delta e^{-\frac{b}{\delta}}\right)\|\nabla v_k(t)\|_2^2 \qquad (6.19)$$

as we have already shown in the proof of Proposition 6.2. Applying the Rayleigh-Ritz formula (6.13), we obtain

$$\|v_k(t)\|_2^2 \le \frac{b^2}{\pi^2}\|\nabla v_k(t)\|_2^2, \quad t > 0. \qquad (6.20)$$

Now, inequalities (6.19) and (6.20) show that

$$\lambda\|v_k(t)\|_2^2 + \nu\|\nabla v_k(t)\|_2^2 + \langle v_3(t)\partial_3 u_b^E, v(t)\rangle > 0$$

for all $t > 0$, whenever condition (6.16) is fulfilled. In this case

$$\frac{\mathrm{d}}{\mathrm{d}t}\|e^{-\lambda t}v_k(t)\|_2^2 < 0, \quad t > 0, \qquad (6.21)$$

holds. Thus, we obtain that

$$\|v_k(t)\|_2 \le e^{\lambda t}\|v_0\|_2, \quad t \ge 0.$$

Now, let $\lambda \ge 0$. By the Cauchy-Schwarz inequality

$$|\langle v_{k3}(t)\partial_3 u_b^E, v_k(t)\rangle| \le \frac{\sqrt{2}\tilde{u}_b}{\delta}\|v_k(t)\|_2^2 \qquad (6.22)$$

holds. From inequalities (6.22) and (6.20) it follows that

$$\lambda\|v_k(t)\|_2^2 + \nu\|\nabla v_k(t)\|_2^2 + \langle v_{k3}(t)\partial_3 u_b^E, v_k(t)\rangle > 0$$

for all $t > 0$, whenever condition (6.17) is fulfilled. The same arguments as above imply that

$$\|v_k(t)\|_2 \le e^{\lambda t}\|v_0\|_2, \quad t \ge 0.$$

By inequality (4.13), we obtain

$$\|v_k(t)\|_2^2 \le \|v_0\|_2^2 \qquad (6.23)$$

for all $t \ge 0$ and every $k \in \mathbb{N}$. By (6.21) and (6.23), these estimates are uniformly in k. Hence, they transfer to the weak solution v, which completes the proof. □

Remark 6.8. Since in the first statement of Theorem 6.7 we assumed $\lambda < 0$, this result implies exponential decay of the weak solution v, if condition (6.16) is satisfied.

6.4 Existence of strong solutions in infinite layers

In this section we will prove the existence of strong solutions with exponential decay to (6.7) for sufficiently small initial values v_0.

The method of our proof needs some preparation including some perturbation theory of sectorial operators. Therefore, we first state an important property of the Stokes operator in $L^p_\sigma(\Omega_b)$.

Proposition 6.9. *Consider the Stokes operator A in $L^p_\sigma(\Omega_b)$. Then A is a sectorial operator of angle $\phi_A = 0$. Furthermore, we have $\sigma(-A) \subset (-\infty, -\frac{\pi^2}{b^2}]$. For $\theta \in (0, \pi)$, let $\lambda \in -\frac{\pi^2}{b^2} + \Sigma_{\pi-\theta}$, and $f \in L^p_\sigma(\Omega_b)$. Then for a solution $v \in D(A)$ of*

$$(\lambda + A)v = f$$

the resolvent estimate

$$\left|\lambda + \frac{\pi^2}{b^2}\right| \|v\|_p + \frac{|\lambda + \frac{\pi^2}{b^2}|}{1 + |\lambda|} \|\nabla^2 v\|_p \leq C_\theta \|f\|_p \tag{6.24}$$

holds. In particular, $-A$ generates an analytic semigroup on $L^p_\sigma(\Omega_b)$ satisfying $\|e^{-tA}\| \leq M e^{-t\frac{\pi^2}{b^2}}$.

For the proof we refer to [AW05].

This enables us to obtain the following statement.

Lemma 6.10. *Let the operator A_{SCE} be given as in (6.9), $\theta \in (0, \pi)$, and let the parameters $\nu, \omega, b, \tilde{u}_b$ fulfill the condition*

$$C_\theta \frac{1}{\nu} \|P\| \frac{b^2}{\pi^2} \left(C\tilde{u}_b + \sqrt{2}\tilde{u}_b + |\omega| \right) < \frac{1}{a}, \tag{6.25}$$

where $a := \sup_{r>0} \|A(r + A)^{-1}\|$. Then A_{SCE} is a sectorial operator of angle $\phi_{A_{SCE}} = \theta$ in $L^p_\sigma(\Omega_b)$. If $\phi_{A_{SCE}} < \frac{\pi}{2}$, then $-A_{SCE}$ generates an analytic semigroup on $L^p_\sigma(\Omega_b)$, which decays exponentially, and $0 \in \rho(A_{SCE})$.

Proof. Looking at the decomposition

$$A_{SCE}v = (\nu A + S + C_{E,1} + C_{E,2})v$$

and Proposition 6.9 we may regard A_{SCE} as a perturbation of the Stokes operator A. Therefore, we will check the conditions of Proposition 1.19. For this purpose, first we have to check the domain of $S+C_{E,1}+C_{E,2}$. Since S and $C_{E,2}$ are bounded operators on $L^p_\sigma(\Omega_b)$ and $C_{E,1}$ is a first order differential operator, we immediately obtain that $D(S + C_{E,1} + C_{E,2}) \supset D(A)$. In the following we will estimate the perturbing operator $S + C_{E,1} + C_{E,2}$ in terms of A. Since for $v \in D(A)$

$$\|(S + C_{E,1} + C_{E,2})v\|_p \leq \|Sv\|_p + \|C_{E,1}v\|_p + \|C_{E,2}v\|_p,$$

we will handle the three terms on the right hand side separately. Since $0 \in \rho(A)$, we have

$$\begin{aligned}\|Sv\|_p &\leq |\omega|\|P\|\|v\|_p \\ &\leq |\omega|\|P\|\|A^{-1}\|\|Av\|_p.\end{aligned}$$

For the second term similar calculations lead to

$$\begin{aligned}\|C_{E,1}v\|_p &\leq \tilde{u}_b\|P\|\|\nabla v\|_p \\ &\leq C\tilde{u}_b\|P\|\|v\|_p^{\frac{1}{2}}\|\Delta v\|_p^{\frac{1}{2}} \\ &\leq C\tilde{u}_b\frac{b}{\pi}\|P\|\|v\|_q^{\frac{1}{2}}\|Av\|_p^{\frac{1}{2}} \\ &\leq C\tilde{u}_b\frac{b}{\pi}\|P\|\|A^{-1}\|^{\frac{1}{2}}\|Av\|_p.\end{aligned}$$

Here we made use of the explicit representation of the Ekman spiral as well as of Proposition 1.40 and the resolvent estimate of the Stokes operator given in (6.24). The constant C derives from the Gagliardo-Nirenberg inequality and the resolvent estimate of the Stokes operator and does not depend on v. Using again the explicit representation of the Ekman spiral, the third term is estimated by

$$\begin{aligned}\|C_{E,2}\|_p &\leq \sqrt{2}\tilde{u}_b\frac{1}{\delta}\|P\|\|v\|_p \\ &\leq \sqrt{2}\tilde{u}_b\frac{1}{\delta}\|P\|\|A^{-1}\|\|Av\|_p.\end{aligned}$$

By Proposition 6.9 we have $\|A^{-1}\| \leq C_\theta\frac{b^2}{\pi^2}$. We set

$$\alpha := C_\theta\frac{1}{\nu}\|P\|\frac{b^2}{\pi^2}\Big(C\tilde{u}_b + \sqrt{2}\tilde{u}_b\frac{1}{\delta} + |\omega|\Big).$$

Then we obtain

$$\|(S + C_{E,1} + C_{E,2})v\|_p \leq \alpha\nu\|Av\|_p.$$

Let $a := \sup_{r>0} \|A(r + A)^{-1}\|$. Thanks to Proposition 6.9, a is finite. By Proposition 1.19, the operator $\nu A + S + C_{E,1} + C_{E,2}$ is sectorial, if

$$\alpha < \frac{1}{a}.$$

Now we take $\mu > 0$, such that

$$\mu < \frac{1}{a} - \alpha.$$

Then

$$\|(S + C_{E,1} + C_{E,2})v - \mu v\|_p \leq (\alpha + \mu)\nu\|Av\|_p,$$

which implies that also the operator $B := \nu A + S + C_{E,1} + C_{E,2} - \mu$ is sectorial. In particular $-B$ generates a bounded analytic semigroup

$$\begin{aligned} e^{-tB} &= e^{-t(\nu A + S + C_{E,1} + C_{E,2} - \mu)} \\ &= e^{-t(\nu A + S + C_{E,1} + C_{E,2})} e^{t\mu} \\ &= e^{-tA_{SCE}} e^{t\mu} \end{aligned}$$

on $L^p_\sigma(\Omega_b)$. Since e^{-tB} is bounded, we obtain

$$\|e^{-tA_{SCE}}\| \leq Ce^{-t\mu},$$

which completes the proof. □

Next, we will prove so-called global L^p-L^q estimates for the semigroup generated by $-A_{SCE}$.

Lemma 6.11. *Assume (6.25) to be satisfied. Let $1 < p \leq q < \infty$. Then there exists $\theta > 0$, such that the semigroup $e^{-tA_{SCE}}$ fulfills the following estimates:*

$$\|e^{-tA_{SCE}}v\|_q \leq C_{p,q} t^{-\frac{3}{2}(\frac{1}{p}-\frac{1}{q})} e^{-t\theta}\|v\|_p, \tag{6.26}$$

$$\|\nabla e^{-tA_{SCE}}v\|_q \leq C_{p,q} t^{-\frac{3}{2}(\frac{1}{p}-\frac{1}{q})-\frac{1}{2}} e^{-t\theta}\|v\|_p. \tag{6.27}$$

Proof. From Lemma 6.10 we know that $-A_{SCE}$ generates a bounded analytic semigroup and that there exists $\mu > 0$, such that $\sigma(-A_{SCE}) \subset \{z \in \mathbb{C} : \operatorname{Re} z \leq -\mu\}$. Take $0 < \theta < \mu$. By standard perturbation theory for analytic semigroups, we obtain that also the shifted operator $-B := -A_{SCE} + \theta$ generates a bounded analytic semigroup $e^{-tB} = e^{-tA_{SCE}}e^{t\theta}$ in $L^p_\sigma(\Omega_b)$ and that $\sigma(-B) \subset \{z \in \mathbb{C} : \operatorname{Re} z \leq \theta - \mu\}$. The boundedness and the analyticity of the semigroup imply

$$\|e^{-tB}f\|_p \leq C\|f\|_p \tag{6.28}$$

and

$$\|Be^{-tB}f\|_p \leq Ct^{-1}\|f\|_p \tag{6.29}$$

for all $f \in L^p_\sigma(\Omega_b)$. Since $0 \in \rho(B)$, we obtain by the open mapping theorem that for all $f \in D(B) = D(A_{SCE})$

$$\|f\|_{2,p} \leq C\|Bf\|_p. \tag{6.30}$$

Inequalities (6.29) and (6.30) together imply

$$\|e^{-tB}f\|_{2,p} \leq C\|Be^{-tB}f\|_p \leq Ct^{-1}\|f\|_p \tag{6.31}$$

for all $f \in L^p_\sigma(\Omega_b)$ and $t > 0$. Proposition 1.40 with $a = \frac{1}{2}$ and inequality (6.31) yield

$$\|\nabla e^{-tB}f\|_p \leq C\|e^{-tB}f\|_{2,p}^{\frac{1}{2}}\|e^{-tB}f\|_p^{\frac{1}{2}} \leq Ct^{-\frac{1}{2}}\|f\|_p \tag{6.32}$$

for all $f \in L^p_\sigma(\Omega_b)$ and $t > 0$. Applying now Proposition 1.40 with $a = 3(\frac{1}{p} - \frac{1}{q})$ and using inequalities (6.28) and (6.32), we obtain the estimate

$$\|e^{-tB}v\|_q \leq C_{p,q}t^{-\frac{3}{2}(\frac{1}{p}-\frac{1}{q})}\|v\|_p \tag{6.33}$$

for all $v \in L^p_\sigma(\Omega_b)$, $t > 0$, and $1 < p \leq q < \infty$. On the other hand, applying Proposition 1.40 with $a = \frac{3}{2}(\frac{1}{p} - \frac{1}{q}) + \frac{1}{2}$ and using inequalities (6.28) and (6.31), we obtain

$$\|\nabla e^{-tB}v\|_r \leq C_{p,q}t^{-\frac{3}{2}(\frac{1}{p}-\frac{1}{q})-\frac{1}{2}}\|v\|_p \tag{6.34}$$

for all $v \in L^p_\sigma(\Omega_b)$, $t > 0$, and $1 < p \leq q < \infty$. Since $e^{-tB} = e^{-tA_{SCE}}e^{t\theta}$, a multiplication of (6.33) and (6.34) with $e^{-t\theta}$ yields the assertion. □

Remark 6.12. Since $L^p_\sigma(\Omega_b)$ is a reflexive Banach space for all $1 < p < \infty$, the adjoint operator A_{SCE}' is densely defined in $L^{p'}_\sigma(\Omega_b)$. In view of Remark 1.15, this implies that A_{SCE}' is a sectorial operator for every $1 < p < \infty$. Moreover, since the spectra $\sigma(A_{SCE})$ and $\sigma(A_{SCE}')$ coincide, we obtain that $-A_{SCE}'$ is the generator of an analytic semigroup $e^{-tA_{SCE}'}$, which is exponentially stable. Hence, by the same arguments as in the proof of Lemma 6.11, the semigroup $e^{-tA_{SCE}'}$ satisfies the estimates (6.26) and (6.27) as well.

The foregoing arguments enable us to show the existence of strong solutions of equations (6.7). To this end, we will make use of Kato's iteration method as described in [Kat84], for which the global L^p-L^q estimates given in Lemma 6.11 are essential.

Theorem 6.13. *Assume (6.25) to be satisfied. Let* $v_0 \in L^3_\sigma(\Omega_b)$*. Then there exists* $T > 0$*, such that system (6.7) possesses a unique strong solution* v *satisfying*

$$v \in C((0,T), D(A_{SCE})) \cap C([0,T), L^3_\sigma(\Omega_b)) \tag{6.35}$$

and

$$t^{\frac{1}{2}} \nabla v \in C([0,T), L^3(\Omega_b)^{3\times 3}). \tag{6.36}$$

If $\|v_0\|_3$ *is sufficiently small, we may choose* $T = \infty$*. In the latter case, we have in addition the stability estimate*

$$\|v(t)\|_3 + t^{\frac{1}{2}} \|\nabla v(t)\|_3 \le C e^{-t\theta} \|v_0\|_3,$$

where $\theta > 0$ *denotes the constant occuring in Lemma 6.11.*

Proof. Following [Wie99], we define an iteration procedure by

$$u_0(t) := e^{-tA_{SCE}} v_0$$

and

$$u_{j+1}(t) := u_0(t) - F(u_j)(t),$$

where

$$F(v)(t) = \int_0^t e^{-(t-s)A_{SCE}} P((v(s) \cdot \nabla) v(s)) \, \mathrm{d}s.$$

For some $\delta \in (0,1)$ and $0 < T \le \infty$ we set

$$\begin{aligned} K_j &:= \sup_{t \le T} e^{t\theta} t^{\frac{1-\delta}{2}} \|u_j(t)\|_{\frac{3}{\delta}}, \\ K'_j &:= \sup_{t \le T} e^{t\theta} t^{\frac{1}{2}} \|\nabla u_j(t)\|_3, \end{aligned}$$

and $R_j := \max\{K_j, K'_j\}$. Note that by the Hölder inequality we have

$$\|P((v \cdot \nabla) v)\|_{\frac{3}{1+\delta}} \le C(\delta) \|v\|_{\frac{3}{\delta}} \|\nabla v\|_3. \tag{6.37}$$

Using (6.26) with $p = \frac{3}{1+\delta}$ and $q = \frac{3}{\delta}$ gives

$$\begin{aligned} K_{j+1} &\le K_0 + \sup_{t \le T} e^{t\theta} t^{\frac{1-\delta}{2}} \int_0^t \|e^{-(t-s)A_{SCE}} P((u_j(s) \cdot \nabla) u_j(s))\|_{\frac{3}{\delta}} \, \mathrm{d}s \\ &\le K_0 + C(\delta) \sup_{t \le T} e^{t\theta} t^{\frac{1-\delta}{2}} \int_0^t (t-s)^{-\frac{1}{2}} \|u_j(s)\|_{\frac{3}{\delta}} \|\nabla u_j(s)\|_3 \, \mathrm{d}s \\ &\le K_0 + C(\delta) \sup_{t \le T} \left\{ e^{t\theta} t^{\frac{1-\delta}{2}} \|u_j(t)\|_{\frac{3}{\delta}} \cdot e^{t\theta} t^{\frac{1}{2}} \|\nabla u_j(t)\|_3 \right\} \\ &\le K_0 + C(\delta) K_j K'_j. \end{aligned}$$

Moreover, since

$$K'_{j+1} \le K'_0 + \sup_{t\le T} e^{t\theta} t^{\frac{1}{2}} \int_0^t \|\nabla e^{-(t-s)A_{SCE}} P((u_j(s)\cdot\nabla)u_j(s))\|_{\frac{3}{\delta}}\, \mathrm{d}s,$$

using (6.27) instead of (6.26) with $p = \frac{3}{1+\delta}$ and $q = 3$, the same calculation yields

$$K'_{j+1} \le K'_0 + C(\delta) K_j K'_j.$$

Together these estimates imply the existence of some $\gamma > 0$, such that

$$R_{j+1} \le R_0 + \gamma R_j^2.$$

Assume now that

$$R_0 \le \frac{1}{6\gamma}. \tag{6.38}$$

Then we obtain by induction that

$$R_j \le 2R_0$$

for all $j \in \mathbb{N}$. For $j \in \mathbb{N}$ consider now the sequence

$$\begin{aligned} w_0(t) &= u_0(t), \\ w_j(t) &= u_j(t) - u_{j-1}(t). \end{aligned}$$

With the same δ and T as in the definition of K_j and K'_j, we define the quantities

$$\begin{aligned} \tilde{K}_j &:= \sup_{t\le T} e^{t\theta} t^{\frac{1-\delta}{2}} \|w_j(t)\|_{\frac{3}{\delta}}, \\ \tilde{K}'_j &:= \sup_{t\le T} e^{t\theta} t^{\frac{1}{2}} \|\nabla w_j(t)\|_3, \end{aligned}$$

as well as $\tilde{R}_j := \max\{\tilde{K}_j, \tilde{K}'_j\}$. Using again (6.37) and (6.26) as well as (6.38), we may now calculate

$$\begin{aligned} \tilde{K}_j &= \sup_{t\le T} e^{t\theta} t^{\frac{1-\delta}{2}} \|u_j(t) - u_{j-1}(t)\|_{\frac{3}{\delta}} \\ &\le C(\delta) \sup_{t\le T} t^{\frac{1-\delta}{2}} \int_0^t (t-s)^{-\frac{1}{2}} \|u_{j-1}(s)\|_{\frac{3}{\delta}} \|\nabla w_j(s)\|_3 \,\mathrm{d}s \\ &\quad + C(\delta) \sup_{t\le T} t^{\frac{1-\delta}{2}} \int_0^t (t-s)^{-\frac{1}{2}} \|w_j(s)\|_{\frac{3}{\delta}} \|\nabla u_j(s)\|_3 \,\mathrm{d}s \\ &\le C(\delta) K_{j-1} \tilde{K}'_j + C(\delta) \tilde{K}_j K'_j \\ &\le C(\delta) \tilde{R}_j (R_j + R_{j-1}) \\ &\le \frac{2}{3} \tilde{R}_j. \end{aligned}$$

Taking again (6.27) instead of (6.26) the same arguments yield

$$\tilde{K}_j' \leq \frac{2}{3}\tilde{R}_j.$$

Hence we obtain the estimate $\tilde{R}_{j+1} \leq \frac{2}{3}\tilde{R}_j$. Since

$$\begin{aligned} \|w_{j+1}(t)\|_3 &\leq C(\delta)\int_0^t (t-s)^{-\frac{\delta}{2}}\|w_j(s)\|_{\frac{3}{\delta}}\|\nabla u_{j-1}(s)\|_3 \,\mathrm{d}s \\ &\quad + C(\delta)\int_0^t (t-s)^{-\frac{\delta}{2}}\|u_j(s)\|_{\frac{3}{\delta}}\|\nabla w_j(s)\|_3 \,\mathrm{d}s \\ &\leq \frac{4}{3}\tilde{R}_j, \end{aligned}$$

we end up with

$$\|w_{j+1}(t)\|_3 \leq C\left(\frac{2}{3}\right)^j. \tag{6.39}$$

In the next step we will show continuity of the iterated functions u_j mapping $[0,T]$ to $L^3_\sigma(\Omega_b)$. For this purpose, we write for $0 < \varepsilon < t_1 < t_2$

$$\begin{aligned} F(u_j)(t_2) - F(u_j)(t_1) &= (e^{-(t_2-t_1)A_{SCE}} - \mathrm{Id}) \\ &\quad \cdot \int_0^{t_1} e^{-(t_1-s)A_{SCE}} P((u_j(s)\cdot\nabla)u_j(s)) \,\mathrm{d}s \\ &\quad + \int_{t_1}^{t_2} e^{-(t_2-s)A_{SCE}} P((u_j(s)\cdot\nabla)u_j(s)) \,\mathrm{d}s \\ &= (e^{-(t_2-t_1)A_{SCE}} - \mathrm{Id})(u_{j+1}(t_1) - u_0(t_1)) + R(t_1,t_2). \end{aligned}$$

Since by (6.26) and (6.37)

$$\begin{aligned} \|R(t_1,t_2)\|_3 &= \left\|\int_{t_1}^{t_2} e^{-(t_2-s)A_{SCE}} P((u_j(s)\cdot\nabla)u_j(s)) \,\mathrm{d}s\right\|_3 \\ &\leq \int_{t_1}^{t_2} C(\delta)(t_2-s)^{-\frac{\delta}{2}}\|u_j(s)\|_{\frac{3}{\delta}}\|\nabla u_j(s)\|_3 \,\mathrm{d}s \\ &\leq \int_{t_1}^{t_2} C(\delta)(t_2-s)^{-\frac{\delta}{2}} s^{-\frac{1}{2}-\frac{\delta}{2}} K_j \cdot s^{-\frac{1}{2}} K_j' \,\mathrm{d}s \\ &\leq C(\delta,\varepsilon)(t_2-t_1)^{1-\frac{\delta}{2}}, \end{aligned}$$

we obtain continuity of u_j for $t > 0$.

For the proof of continuity at $t = 0$, for $\mu > 0$ we define

$$E_j := \sup_{0\leq t\leq\varepsilon} \|u_j(t) - e^{-\mu A_{SCE}} v_0\|_3.$$

Then we estimate

$$\begin{aligned}
E_{j+1} &\leq \sup_{0\leq t\leq \varepsilon} \|e^{-tA_{SCE}}v_0 - e^{-\mu A_{SCE}}v_0\|_3 \\
&\quad + \sup_{0\leq t\leq \varepsilon} \left\| \int_0^t e^{-(t-s)A_{SCE}} P((u_j(s)\cdot\nabla)u_j(s))\, ds \right\|_3 \\
&\leq E_0 + \sup_{0\leq t\leq \varepsilon} \int_0^t \|e^{-(t-s)A_{SCE}} P\big(((u_j(s) - e^{-\mu A_{SCE}}v_0)\cdot\nabla)u_j(s)\big)\|_3 \\
&\quad + \|e^{-(t-s)A_{SCE}} P((e^{-\mu A_{SCE}}v_0\cdot\nabla)u_j(s))\|_3 \, ds \\
&\leq E_0 + \gamma E_j K_j' + C(\delta) \sup_{0\leq t\leq \varepsilon} \int_0^t (t-s)^{-\frac{\delta}{2}} s^{-\frac{1}{2}} K_j' \|e^{-\mu A_{SCE}}v_0\|_{\frac{3}{\delta}} \, ds \\
&\leq E_0 + \frac{1}{3}E_j + C(\delta)\|v_0\|_3 \left(\frac{\varepsilon}{\mu}\right)^{\frac{1-\delta}{2}}.
\end{aligned}$$

By induction we see that

$$E_j \leq 2E_0 + 2C(\delta)\|v_0\|_3 \left(\frac{\varepsilon}{\mu}\right)^{\frac{1-\delta}{2}}.$$

Let $\tau > 0$. Since

$$\sup_{0\leq t\leq \varepsilon} \|u_j(t) - v_0\|_3 \leq \sup_{0\leq t\leq \varepsilon} \|u_j(t) - e^{-\mu A_{SCE}}v_0\|_3 + \sup_{0\leq t\leq \varepsilon} \|e^{-\mu A_{SCE}}v_0 - v_0\|_3,$$

we now may choose first μ and then ε, such that

$$\sup_{0\leq t\leq \varepsilon} \|u_j(t) - v_0\|_3 \leq \tau,$$

which implies continuity of u_j at $t = 0$. Hence we have $u_j \in C([0,T], L^3_\sigma(\Omega_b))$. In view of (6.39) this proves that $u_j(t) = \sum_{k=0}^{j} w_k(t)$ is a Cauchy sequence in $C([0,T], L^3_\sigma(\Omega_b))$ converging to a solution $u \in C([0,T], L^3_\sigma(\Omega_b))$ of the integral equation

$$u(t) = e^{-tA_{SCE}}v_0 - \int_0^t e^{-(t-s)A_{SCE}} P((u(s)\cdot\nabla)u(s))\, ds. \tag{6.40}$$

That means u is a mild solution of (6.8) fulfilling the estimates

$$e^{t\theta} t^{\frac{1-\delta}{2}} \|u(t)\|_{\frac{3}{\delta}} \leq 2R_0, \tag{6.41}$$

$$e^{t\theta} t^{\frac{1}{2}} \|\nabla u(t)\|_3 \leq 2R_0, \tag{6.42}$$

for $t \leq T$.

In the following we will show that the mild solution u is in fact a strong solution of (6.7) for $0 < t < T$. For this purpose we take $0 < \varepsilon < \delta$ and consider

$$\begin{aligned} S_j &= \sup_{0<h,t+h\leq T} t^{\frac{1-\varepsilon}{2}} h^{-\frac{\delta-\varepsilon}{2}} \|u_j(t+h) - u_j(t)\|_{\frac{3}{\delta}}, \\ S_j' &= \sup_{0<h,t+h\leq T} t^{\frac{1+\delta-\varepsilon}{2}} h^{-\frac{\delta-\varepsilon}{2}} \|\nabla u_j(t+h) - \nabla u_j(t)\|_3, \end{aligned}$$

and $\tilde{S}_j = \max\{S_j, S_j'\}$. Now let $t \geq 2h$. Then we calculate

$$\begin{aligned} \|u_0(t+h) - u_0(t)\|_{\frac{3}{\delta}} &= \|e^{-(t+h)A_{SCE}} v_0 - e^{-tA_{SCE}} v_0\|_{\frac{3}{\delta}} \\ &= \|A_{SCE} \int_0^h e^{-sA_{SCE}} e^{-tA_{SCE}} v_0 \, \mathrm{d}s\|_{\frac{3}{\delta}} \\ &\leq Ch\|A_{SCE} e^{-tA_{SCE}} v_0\|_{\frac{3}{\delta}} \\ &\leq C2\frac{h}{t}\|e^{-\frac{t}{2}A_{SCE}} v_0\|_{\frac{3}{\delta}} \\ &= C\frac{h^{(\delta-\varepsilon)/2}}{t^{(1-\varepsilon)/2}} R_0. \end{aligned}$$

In the case $t \leq 2h$ we may write $t = \kappa h$ with $0 < \kappa \leq 2$. Then we have

$$\begin{aligned} & t^{\frac{1-\varepsilon}{2}} h^{-\frac{\delta-\varepsilon}{2}} \|e^{-(t+h)A_{SCE}} v_0 - e^{-tA_{SCE}} v_0\|_{\frac{3}{\delta}} \\ = {} & \kappa^{\frac{1-\varepsilon}{2}} h^{\frac{1-\delta}{2}} \|e^{-(1+\kappa)hA_{SCE}} v_0 - e^{-\kappa h A_{SCE}} v_0\|_{\frac{3}{\delta}} \\ \leq {} & ((1+\kappa)^{-\frac{1-\delta}{2}} + \kappa^{-\frac{1-\delta}{2}}) \kappa^{\frac{1-\varepsilon}{2}} R_0. \end{aligned}$$

So we get $S_0 \leq CR_0$. On the other hand we have by the analyticity of $e^{-tA_{SCE}}$ the inequality $\|A_{SCE} e^{-tA_{SCE}}\| \leq Ct^{-\frac{1}{2}}\|\nabla e^{-tA_{SCE}}\|$. This can be seen as follows: since $e^{-tA_{SCE}}$ is an analytic semigroup, Remark 1.15 implies $\|A_{SCE} e^{-2tA_{SCE}} f\|_3 \leq C\frac{1}{t}\|e^{-tA_{SCE}} f\|_3$ for $f \in L^3_\sigma(\Omega_b)$. Moreover, we have

$$\begin{aligned} \|A_{SCE} e^{-2tA_{SCE}} f\|_3 &= \|e^{-tA_{SCE}} A_{SCE} e^{-tA_{SCE}} f\|_3 \\ &\leq C\|A_{SCE} e^{-tA_{SCE}} f\|_3 \\ &\leq C\|e^{-tA_{SCE}} f\|_{2,3}. \end{aligned}$$

By interpolation we obtain

$$\begin{aligned} \|A_{SCE} e^{-tA_{SCE}} f\|_3 &\leq C\|A_{SCE} e^{-2tA_{SCE}} f\|_3 \\ &\leq Ct^{-\frac{1}{2}}\|e^{-tA_{SCE}} f\|_{1,3}. \end{aligned}$$

Now, Poincaré's inequality implies $\|A_{SCE} e^{-tA_{SCE}} f\|_3 \leq Ct^{-\frac{1}{2}}\|\nabla e^{-tA_{SCE}} f\|_3$.

Thus, we have for $t \geq 2h$

$$\begin{aligned}
\|\nabla u_0(t+h) - \nabla u_0(t)\|_3 &= \|\nabla(e^{-(t+h)A_{SCE}}v_0 - e^{-tA_{SCE}}v_0)\|_3 \\
&\leq C2^{\frac{1}{2}}t^{-\frac{1}{2}}\|(e^{-hA_{SCE}} - \mathrm{Id})e^{-\frac{t}{2}A_{SCE}}v_0\|_3 \\
&\leq C2^{\frac{1}{2}}t^{-\frac{1}{2}}h\|A_{SCE}e^{-\frac{t}{2}A_{SCE}}v_0\|_3 \\
&\leq C2\frac{h}{t}\|\nabla e^{-\frac{t}{2}A_{SCE}}v_0\|_3 \\
&\leq C\frac{h}{t^{-3/2}}R_0 \\
&\leq C\frac{h^{(\delta-\varepsilon)/2}}{t^{(1+\delta-\varepsilon)/2}}R_0.
\end{aligned}$$

Furthermore, for $t = \kappa h$ with $0 < \kappa \leq 2$ we obtain

$$\begin{aligned}
& t^{\frac{1+\delta-\varepsilon}{2}}h^{-\frac{\delta-\varepsilon}{2}}\|\nabla e^{-(t+h)A_{SCE}}v_0 - \nabla e^{-tA_{SCE}}v_0\|_3 \\
= {} & \kappa^{\frac{1+\delta-\varepsilon}{2}}h^{\frac{1}{2}}\|\nabla e^{-(1+\kappa)hA_{SCE}}v_0 - \nabla e^{-\kappa hA_{SCE}}v_0\|_3 \\
\leq {} & ((1+\kappa)^{-\frac{1}{2}} + \kappa^{-\frac{1}{2}})\kappa^{\frac{1+\delta-\varepsilon}{2}}R_0.
\end{aligned}$$

Hence $S_0' \leq CR_0$, too. This implies the inequality

$$\tilde{S}_{j+1} \leq \tilde{S}_0 + CR_0 + C_\varepsilon \tilde{S}_j R_0.$$

Assume now

$$C_\varepsilon R_0 \leq \frac{1}{2}. \tag{6.43}$$

By the results of the calculations above, S_0 may be estimated in terms of R_0. Hence, we obtain $\tilde{S}_j \leq CR_0$ for all $j \in \mathbb{N}$. Since these estimates are uniform in j, they transfer to the limit solution u. Having this at hand we can show for $G(s) = P(u(s)\cdot\nabla)u(s)$ that

$$\begin{aligned}
& \sup_{0<h,t+h\leq T} \frac{\|G(s+h) - G(s)\|_{\frac{3}{1+\delta}}}{h^{(\delta-\epsilon)/2}} \\
\leq {} & C(\delta) \sup_{0<h,t+h\leq T} \frac{\|((u(s+h) - u(s))\cdot\nabla)u(s+h)\|_{\frac{3}{1+\delta}}}{h^{(\delta-\epsilon)/2}} \\
& + C(\delta) \sup_{0<h,t+h\leq T} \frac{\|(u(s)\cdot\nabla)(u(s+h) - u(s))\|_{\frac{3}{1+\delta}}}{h^{(\delta-\epsilon)/2}} \\
\leq {} & C(\delta)R_0 \sup_{0<h,t+h\leq T} t^{-\frac{1}{2}}\frac{\|u(s+h) - u(s)\|_{\frac{3}{\delta}}}{h^{(\delta-\epsilon)/2}} \\
& + C(\delta)R_0 \sup_{0<h,t+h\leq T} t^{-\frac{1-\delta}{2}}\frac{\|\nabla(u(s+h) - u(s))\|_3}{h^{(\delta-\epsilon)/2}},
\end{aligned}$$

which implies that for every $0 < t_0 < T$ the nonlinear term $G : [t_0, T) \to L^{\frac{3}{1+\delta}}(\Omega_b)^3$ is $\frac{\delta-\varepsilon}{2}$-Hölder continuous.

Next, note that $N_\rho(u)(t) = \int_0^{t-\rho} e^{-(t-s)A_{SCE}}G(s)\,\mathrm{d}s \in D(A_{SCE})$ for $t > t_0$ and $\rho < t - t_0$. Since

$$\begin{aligned} A_{SCE}N_\rho(t) &= \int_0^{t_0} A_{SCE}e^{-(t-s)A_{SCE}}G(s)\,\mathrm{d}s \\ &\quad + \int_{t_0}^{t-\rho} A_{SCE}e^{-(t-s)A_{SCE}}(G(s) - G(t))\,\mathrm{d}s \\ &\quad - e^{-\rho A_{SCE}}G(t) + e^{-(t-t_0)A_{SCE}}G(t), \end{aligned}$$

we see that N_ρ converges in $C((t_0,T), L^{\frac{3}{1+\delta}}(\Omega_b))$ for $\rho \to 0$. Hence it follows that $u(t) = e^{-tA_{SCE}}v_0 + N(u)(t) \in D(A_{SCE}) \subset W^{2,\frac{3}{1+\delta}}(\Omega_b)^3$ for $t \in (0,T)$. Since we work in a three dimensional domain, by Sobolev's embedding we obtain that $u(t) \in L^\infty(\Omega_b)^3$. Moreover, $\nabla u(t) \in L^3(\Omega_b)^{3\times 3}$, which implies $G(t) \in L^3(\Omega_b)^3$. It follows that $u(t) \in D(A_{SCE})$. This yields $A_{SCE}u \in C((0,T), L^3_\sigma(\Omega_b))$ and $\partial_t u + A_{SCE}u + P(u\cdot\nabla)u = 0$. Defining the pressure p by $\nabla p = (\mathrm{Id} - P)(\Delta u - (u\cdot\nabla)u)$, we have constructed a complete solution of (6.7).

For the existence of a strong solution the assumptions (6.38) and (6.43) are essential. Since

$$R_0 \le C\|v_0\|_3,$$

these assumptions are always fulfilled in case $\|v_0\|_3$ is sufficiently small. Then we may choose $T = \infty$ and obtain a global strong solution. Otherwise, observe that for $\mu > 0$

$$\begin{aligned} K_0 &\le \sup_{0\le t\le T} t^{\frac{1-\delta}{2}}\|e^{-tA_{SCE}}(\mathrm{Id} - e^{-\mu A_{SCE}})v_0\|_{\frac{3}{\delta}} \\ &\quad + \sup_{0\le t\le T} t^{\frac{1-\delta}{2}}\|e^{-(t+\mu)A_{SCE}}v_0\|_{\frac{3}{\delta}} \\ &\le C\|v_0 - e^{-\mu A_{SCE}}v_0\|_3 + C\left(\frac{T}{T+\mu}\right)^{\frac{1-\delta}{2}}\|v_0\|_3, \end{aligned}$$

where we made use of (6.26) again. The same technique with (6.27) yields

$$\begin{aligned} K_0' &\le \sup_{0\le t\le T} t^{\frac{1}{2}}\|\nabla e^{-tA_{SCE}}(\mathrm{Id} - e^{-\mu A_{SCE}})v_0\|_3 \\ &\quad + \sup_{0\le t\le T} t^{\frac{1}{2}}\|\nabla e^{-(t+\mu)A_{SCE}}v_0\|_3 \\ &\le C\|v_0 - e^{-\mu A_{SCE}}v_0\|_3 + C\left(\frac{T}{T+\mu}\right)^{\frac{1}{2}}\|v_0\|_3. \end{aligned}$$

Since $0 < \delta < 1$, we have

$$\left(\frac{T}{T+\mu}\right)^{\frac{1}{2}} \le \left(\frac{T}{T+\mu}\right)^{\frac{1-\delta}{2}}.$$

Hence we obtain

$$R_0 \leq C\|v_0 - e^{-\mu A_{SCE}} v_0\|_3 + C\left(\frac{T}{T+\mu}\right)^{\frac{1-\delta}{2}} \|v_0\|_3.$$

Therefore, for every $\alpha > 0$ we may choose first μ and then T, such that $R_0 \leq \alpha$. This proves that there exists a local solution of (6.7) for every $v_0 \in L^3_\sigma(\Omega_b)$.

It remains to prove the uniqueness of the solution v. For this purpose, let $q > 3$ and consider the quantity

$$L_j := \sup_{t \leq T} t^\beta \|u_j(t)\|_q,$$

where $\beta := \frac{3}{2}\left(\frac{1}{3} - \frac{1}{q}\right)$. Using the estimates given in Lemma 6.11, we obtain by the same calculations as above

$$L_{j+1} \leq K_0 + C_q L_j K_j'.$$

By induction, this yields that there exists a constant $\tilde{L}$, such that $L_j \leq \tilde{L}$ for all $j \in \mathbb{N}$. Similarly as above, we may show that the functions $t^\beta u_j(t)$ are continuous from $[0,T)$ to $L^q_\sigma(\Omega_b)$ and that the sequence $\left(t^\beta u_j(t)\right)_{j\in\mathbb{N}}$ converges uniformly. Due to the analyticity of the semigroup $e^{-tA_{SCE}}$ and Lemma 6.11, we obtain that $\lim_{t\to 0} t^\beta \|u_0(t)\|_q = 0$. Moreover, using Lemma 6.11 we have

$$\|u_j(t)\|_q \leq \|u_0(t)\|_q + C\tilde{L}R_0 \int_0^t (t-s)^{-\frac{1}{2}} s^{-\beta}\, ds \leq Ct^{-\beta}.$$

Hence, the limit function v satisfies

$$t^\beta v \in C([0,T), L^q_\sigma(\Omega_b))$$

and

$$\lim_{t\to 0} t^\beta \|v(t)\|_q = 0.$$

Now, the uniqueness of the strong solution v on $[0,T)$ follows directly from [Gig86, Theorem 1], where we have to choose $\alpha = \gamma = 1$ and $m = 2$. □

Bibliography

[Abe05a] H. Abels, *Reduced and generalized Stokes resolvent equations in asymptotically flat layers. I. Unique solvability*, J. Math. Fluid Mech. **7** (2005), 201–222.

[Abe05b] ———, *Reduced and generalized Stokes resolvent equations in asymptotically flat layers. II. H_∞-calculus*, J. Math. Fluid Mech. **7** (2005), 223–260.

[ABHN01] W. Arendt, C. J. K. Batty, M. Hieber, and F. Neubrander, *Vector-valued Laplace transforms and Cauchy problems*, Monographs in Mathematics, vol. 96, Birkhäuser Verlag, Basel, 2001.

[ADN59] S. Agmon, A. Douglis, and L. Nirenberg, *Estimates near the boundary for solutions of elliptic partial differential equations satisfying general boundary conditions. I*, Comm. Pure Appl. Math. **12** (1959), 623–727.

[Ama95] H. Amann, *Linear and Quasilinear Parabolic Problems. Vol. I*, Monographs in Mathematics, vol. 89, Birkhäuser Boston Inc., Boston, MA, 1995.

[Are94] W. Arendt, *Gaussian estimates and interpolation of the spectrum in L^p*, Differential Integral Equations **7** (1994), 1153–1168.

[AS03a] T. Abe and Y. Shibata, *On a resolvent estimate of the Stokes equation on an infinite layer*, J. Math. Soc. Japan **55** (2003), 469–497.

[AS03b] ———, *On a resolvent estimate of the Stokes equation on an infinite layer. II. $\lambda = 0$ case*, J. Math. Fluid Mech. **5** (2003), 245–274.

[AW05] H. Abels and M. Wiegner, *Resolvent estimates for the Stokes operator on an infinite layer*, Differential Integral Equations **18** (2005), 1081–1110.

[BM92] W. Borchers and T. Miyakawa, *L^2-decay for Navier-Stokes flows in unbounded domains, with application to exterior stationary flows*, Arch. Rational Mech. Anal. **118** (1992), 273–295.

[BM95] ______, *On stability of exterior stationary Navier-Stokes flows*, Acta Math. **174** (1995), 311–382.

[BMN01] A. Babin, A. Mahalov, and B. Nicolaenko, *3D Navier-Stokes and Euler equations with initial data characterized by uniformly large vorticity*, Indiana Univ. Math. J. **50** (2001), 1–35.

[Bog79] M. E. Bogovskiĭ, *Solution of the first boundary value problem for an equation of continuity of an incompressible medium*, Dokl. Akad. Nauk SSSR **248** (1979), 1037–1040.

[BS90] W. Borchers and H. Sohr, *On the equations* $\operatorname{rot} \mathbf{v} = \mathbf{g}$ *and* $\operatorname{div} \mathbf{u} = f$ *with zero boundary conditions*, Hokkaido Math. J. **19** (1990), 67–87.

[Chu92] S.-K. Chua, *Extension theorems on weighted Sobolev spaces*, Indiana Univ. Math. J. **41** (1992), 1027–1076.

[DDG99] B. Desjardins, E. Dormy, and E. Grenier, *Stability of mixed Ekman-Hartmann boundary layers*, Nonlinearity **12** (1999), 181–199.

[DG03] B. Desjardins and E. Grenier, *Linear instability implies nonlinear instability for various types of viscous boundary layers*, Ann. Inst. H. Poincaré Anal. Non Linéaire **20** (2003), 87–106.

[DHP01] W. Desch, M. Hieber, and J. Prüss, *L^p-theory of the Stokes equation in a half space*, J. Evol. Equ. **1** (2001), 115–142.

[DHP03] R. Denk, M. Hieber, and J. Prüss, *$\mathcal{R}$-boundedness, Fourier multipliers and problems of elliptic and parabolic type*, vol. 788, Mem. Amer. Math. Soc., 2003.

[Dor93] G. Dore, *L^p regularity for abstract differential equations*, In: Functional analysis and related topics, 1991 (Kyoto), Lecture Notes in Math., vol. 1540, Springer, Berlin, 1993, pp. 25–38.

[Ekm05] V. W. Ekman, *On the influence of the earth's rotation on ocean currents*, Arkiv Matem. Astr. Fysik (Stockholm) **11** (1905), 1–52.

[EN00] K.-J. Engel and R. Nagel, *One-parameter Semigroups for Linear Evolution Equations*, Springer-Verlag, New York, 2000.

[Far03] R. Farwig, *Weighted L^q-Helmholtz decompositions in infinite cylinders and in infinite layers*, Adv. Differential Equations **8** (2003), 357–384.

[FK64] H. Fujita and T. Kato, *On the Navier-Stokes initial value problem. I*, Arch. Rational Mech. Anal. **16** (1964), 269–315.

[Fri69] A. Friedman, *Partial Differential Equations*, Holt, Rinehart and Winston, Inc., New York, 1969.

[Frö07] A. Fröhlich, *The Stokes operator in weighted L^q-spaces. II. Weighted resolvent estimates and maximal L^p-regularity*, Math. Ann. **339** (2007), 287–316.

[Gal94a] G. P. Galdi, *An introduction to the mathematical theory of the Navier-Stokes equations. Vol. I*, Springer Tracts in Natural Philosophy, vol. 38, Springer-Verlag, New York, 1994.

[Gal94b] ______, *An introduction to the mathematical theory of the Navier-Stokes equations. Vol. II*, Springer Tracts in Natural Philosophy, vol. 39, Springer-Verlag, New York, 1994.

[GHH06a] M. Geissert, H. Heck, and M. Hieber, *L^p-theory of the Navier-Stokes flow in the exterior of a moving or rotating obstacle*, J. Reine Angew. Math. **596** (2006), 45–62.

[GHH06b] M. Geißert, H. Heck, and M. Hieber, *On the equation* $\operatorname{div} u = g$ *and Bogovskiĭ's operator in Sobolev spaces of negative order*, In: Partial differential equations and functional analysis, Oper. Theory Adv. Appl., vol. 168, Birkhäuser, Basel, 2006, pp. 113–121.

[GHH+08] M. Geissert, M. Hess, M. Hieber, C. Schwarz, and K. Stavrakidis, *Maximal $L^p - L^q$-estimates for the Stokes-equation: A short proof of Solonnikov's theorem*, J. Math. Fluid Mech. **11** (2008), 1–14.

[Gig86] Y. Giga, *Solutions for semilinear parabolic equations in L^p and regularity of weak solutions of the Navier-Stokes system*, J. Differential Equations **62** (1986), 186–212.

[GIM+07] Y. Giga, K. Inui, A. Mahalov, S. Matsui, and J. Saal, *Rotating Navier-Stokes equations in $\mathbb{R}^3_+$ with initial data nondecreasing at infinity: the Ekman boundary layer problem*, Arch. Ration. Mech. Anal. **186** (2007), 177–224.

[Gre80] H. P. Greenspan, *The Theory of Rotating Fluids*, Cambridge University Press, Cambridge, 1980.

[GS91a] Y. Giga and H. Sohr, *Abstract L^p estimates for the Cauchy problem with applications to the Navier-Stokes equations in exterior domains*, J. Funct. Anal. **102** (1991), 72–94.

[GS91b] G. Grubb and V. A. Solonnikov, *Boundary value problems for the nonstationary Navier-Stokes equations treated by pseudo-differential methods*, Math. Scand. **69** (1991), 217–290 (1992).

[Jon81] P. W. Jones, *Quasiconformal mappings and extendability of functions in Sobolev spaces*, Acta Math. **147** (1981), 71–88.

[Kat84] T. Kato, *Strong L^p-solutions of the Navier-Stokes equation in $\mathbf{R}^m$, with applications to weak solutions*, Math. Z. **187** (1984), 471–480.

[KL63] A. A. Kiselev and O. A. Ladyženskaya, *On the existence and uniqueness of solutionis of the non-stationary problem for flows non-compressible fluids*, Amer. Math. Soc. Transl. Ser. **24** (1963), 79–106.

[KW04] P. C. Kunstmann and L. Weis, *Maximal L_p-regularity for parabolic equations, Fourier multiplier theorems and H^∞-functional calculus*, In: Functional analytic methods for evolution equations, Lecture Notes in Math., vol. 1855, Springer, Berlin, 2004, pp. 65–311.

[Ler34] J. Leray, *Sur le mouvement d'un liquide visqueux emplissant l'espace*, Acta Math. **63** (1934), 193–248.

[Lil66] D. K. Lilly, *On the instability of Ekman boundary flow*, J. Atmos. Sci. **23** (1966), 481–494.

[Maz85] V. G. Maz'ja, *Sobolev Spaces*, Springer Series in Soviet Mathematics, Springer-Verlag, Berlin, 1985.

[McC81] M. McCracken, *The resolvent problem for the Stokes equations on halfspace in L_p*, SIAM J. Math. Anal. **12** (1981), 201–228.

[Miy94] T. Miyakawa, *The Helmholtz decomposition of vector fields in some unbounded domains*, Math. J. Toyama Univ. **17** (1994), 115–149.

[MS88] T. Miyakawa and H. Sohr, *On energy inequality, smoothness and large time behavior in L^2 for weak solutions of the Navier-Stokes equations in exterior domains*, Math. Z. **199** (1988), 455–478.

[NS03] A. Noll and J. Saal, *H^∞-calculus for the Stokes operator on L_q-spaces*, Math. Z. **244** (2003), 651–688.

[Sob64] P. E. Sobolevskiĭ, *Coerciveness inequalities for abstract parabolic equations*, Dokl. Akad. Nauk SSSR **157** (1964), 52–55.

[Soh01] Hermann Sohr, *The Navier-Stokes equations*, Birkhäuser Verlag, Basel, 2001.

[Sol77] V. A. Solonnikov, *Estimates for solutions of nonstationary Navier-Stokes equations*, J. Soviet Math. **8** (1977), 467–523.

[SS03] Y. Shibata and S. Shimizu, *On a resolvent estimate for the Stokes system with Neumann boundary condition*, Differential Integral Equations **16** (2003), 385–426.

[SW71] E. M. Stein and G. Weiss, *Introduction to Fourier Analysis on Euclidean Spaces*, Princeton University Press, Princeton, N.J., 1971.

[TCY05] W. Tang, C. P. Caulfield, and W. R. Young, *Bounds on dissipation in stress-driven flow in a rotating frame*, J. Fluid Mech. **540** (2005), 373–391.

[Tem77] R. Temam, *Navier-Stokes Equations. Theory and Numerical Analysis*, North-Holland Publishing Co., Amsterdam, 1977.

[Tri95] H. Triebel, *Interpolation Theory, Function Spaces, Differential Operators*, 2nd ed., Johann Ambrosius Barth, Heidelberg, Leipzig, 1995.

[Uka87] S. Ukai, *A solution formula for the Stokes equation in* $\mathbb{R}^n_+$, Comm. Pure Appl. Math. **40** (1987), 611–621.

[Wei01] L. Weis, *Operator-valued Fourier multiplier theorems and maximal* L_p*-regularity*, Math. Ann. **319** (2001), 735–758.

[Wie99] M. Wiegner, *The Navier-Stokes equations – a neverending challenge?*, Jahresber. Deutsch. Math.-Verein. **101** (1999), 1–25.